Manufacturing Parameters and Entrepreneurship

Manufacturing Parameters and Entrepreneurship

Space Consideration

Sanjay Sharma

CRC Press
Taylor & Francis Group
Boca Raton London New York

CRC Press is an imprint of the
Taylor & Francis Group, an **informa** business

First edition published 2021

by CRC Press
6000 Broken Sound Parkway NW, Suite 300, Boca Raton, FL 33487-2742

and by CRC Press
2 Park Square, Milton Park, Abingdon, Oxon, OX14 4RN

© 2021 Taylor & Francis Group, LLC

CRC Press is an imprint of Taylor & Francis Group, LLC

ISBN: 9780367558543 (hbk)
ISBN: 9781003107163 (ebk)

Typeset in Times LT Std and Optima LT Std
by KnowledgeWorks Global Ltd.

Contents

Preface

Entrepreneurship has many facets including commercial and technical aspects, among others. Manufacturing entrepreneurship may specifically relate to the production or manufacturing of items. After conceptualizing a product or an item, suitable space is needed for an arrangement of facilities and storage. While manufacturing at the available space, quality aspects need to be considered along with the production backorders. This book is concerned with manufacturing entrepreneurship while taking into consideration various factors/parameters and space requirement. Success in the context of manufacturing entrepreneurship lies in knowing the operational factors and their challenge in the real world. The concepts generated are useful for the new production setup as well as for a running organization. After the start of a company, manufacturing entrepreneurship has to deal with uncertainties in which an operational factor or manufacturing parameter can change. Under such a scenario, a suitable response is necessary to obtain the overall beneficial manufacturing environment or to make it feasible. This is also helpful in facilitating the change/ disturbance or minimizing its ill effects. The present book is especially concerned with manufacturing entrepreneurship in the context of space consideration.

Dr. Sanjay Sharma
National Institute of Industrial Engineering

About the Author

Dr. Sanjay Sharma is Professor at National Institute of Industrial Engineering (NITIE), Mumbai, India. He is an operations and supply chain management educator and researcher. He has more than three decades of experience in industrial, managerial, teaching/training, consultancy, and research; he also has many awards/honors to his credit. He has published six books and papers in various journals such as *The European Journal of Operational Research, International Journal of Production Economics, Computers & Operations Research, International Journal of Advanced Manufacturing Technology, Journal of the Operational Research Society,* and *Computers and Industrial Engineering.* Sharma is also a reviewer for several international journals, and is also on the editorial board of a few journals, including the *International Journal of Logistics Management* (an 'A' rating Emerald Journal).

1 Introduction

Entrepreneurship has many facets including commercial and technical aspects (Figure 1.1) among others. Some examples are:

   i. Idea generation for production/services enterprise
   ii. Selection of location for manufacture/service provision
   iii. Feasibility study
   iv. Layout for production/service provision
   v. Technology/facility selection and installation
   vi. Necessary information technology/system support
   vii. Product/process cost estimate
  viii. Price determination for product/services
   ix. Frequent planning with respect to operational factors fluctuation, in order to become viable and also to survive in the competitive world.

Entrepreneurship deals with manufacturing and services (Figure 1.2). Manufacturing pertains to the production of goods/products, whereas services may not be specifically dedicated to production activities including machining.

The services might include, among others:

   i. Accounts
   ii. Administrative
   iii. Transportation
   iv. Personnel/recruitment
   v. Travel
   vi. Hospitality
   vii. Financial
  viii. Hospital/medicine

Some of the services might be associated with the production organization also. However, as this book deals with manufacturing entrepreneurship, the specific issues (in the context of manufacture) are as follows:

   a. Identification of potential customer group
   b. Identification of appropriate manufactured product and its design
   c. Suitable plant location and production facility layout
   d. Price determination for an offered product on the basis of total incurred cost estimate
   e. A good idea of the operational factors and their effect on the relevant costs

Manufacturing entrepreneurship may specifically relate to the production or manufacture of items. After conceptualizing a product or an item, suitable space is needed

1

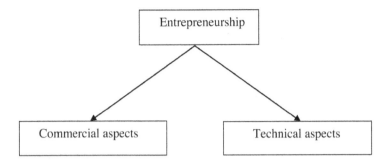

FIGURE 1.1   Aspects of entrepreneurship.

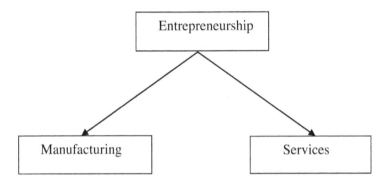

FIGURE 1.2   Manufacturing/services entrepreneurship.

for the arrangement of facilities and storage. While manufacturing at the available space, quality aspects need to be considered along with production backorders. Success in the context of manufacturing entrepreneurship lies in knowing the operational factors or manufacturing parameters and their challenge in the real world. The concepts generated are useful in a new production setup as well as a running organization. After the start of a company, manufacturing entrepreneurship has to deal with uncertainties in which an operational factor or manufacturing parameter can change. In such a scenario, a suitable response is necessary to create an overall beneficial manufacturing environment or to make it feasible. This is also helpful in facilitating the change/disturbance or minimizing its ill effects. This book is especially concerned with manufacturing entrepreneurship in the context of space consideration.

1.1 MANUFACTURING SPACE

In general, suitable space needs to be provided for keeping the:

    a. Raw material/input material
    b. Work-in-process inventory
    c. Final product/assembly

1.1.1 RAW MATERIAL/INPUT MATERIAL

For example, hot-rolled mild steel sheets can be a significant raw material for the sheet metal fabrication industry. As shown in Figure 1.3, this input material is released to the first facility (usually a shearing machine). After release of the material, it is stacked one over the other, on one side of the facility (i.e., the input side). Length and width (e.g., 2500 × 1250 mm) of the sheet metal forms the basis for selecting a suitable storage space. However, some additional space is needed for handling the material, i.e., in order to take each steel sheet to the shearing machine to cut it to make square sheet metal pieces of equal length and width (i.e., 625 × 625 mm).

1.1.2 WORK-IN-PROCESS INVENTORY

Work-in-process inventory is at the intermediate stage of the production. For example, square sheet metal pieces of equal length and width (i.e., 625 × 625 mm) are available on

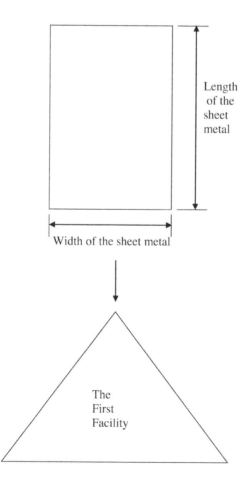

FIGURE 1.3 Release of raw material to the first facility.

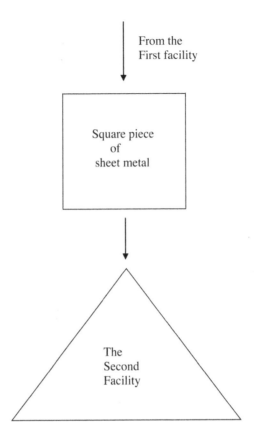

FIGURE 1.4   Work-in-process inventory (between the first and the second facility).

the output side of the first facility, i.e., the shearing machine. These are fed to the second facility, i.e., the circle cutting machine. The role of the circle cutting machine is to convert the square pieces into circular pieces of approximately 615 mm diameter for further processing in order to fabricate specific products depending on the design and specifications of that particular product. As shown in Figure 1.4, square sheet metal pieces (that are stacked one over the other) are on the input side of the second facility, i.e., the circle cutting machine along with the provision of suitable space for such work-in-process inventory. Dimension of the square sheet metal forms the basis for space requirement along with the provision for handling the material, i.e., for taking it to the second facility.

1.1.3   FINAL PRODUCT/ASSEMBLY

Final assembly or final product is available at the output side of the last facility as shown in Figure 1.5. Depending on the design, shape, and size of the final product, these might not be stacked one over the other in larger numbers. Area occupied by the product on the floor forms the basis for overall space requirement considering other factors also. Adjustments are to be made depending on the specific products. For example, pipes are manufactured in certain diameter, thickness, and length.

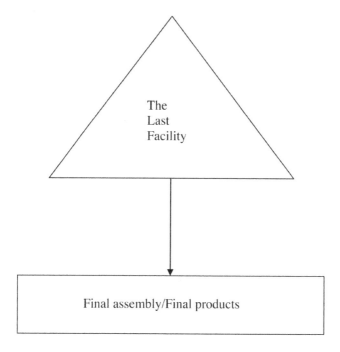

FIGURE 1.5 Final products/assembly at the output side of the last facility.

Few hpipes or tubes of similar specification may be clustered together at the output side of the tube mill. This may be needed to create appropriate lot for further handling with the help of crane or subsequent necessary processing/inspection depending on the precise customer requirement. However, for heavy equipment assembly like transformer assembly, area covered by one assembly is relatively larger and therefore leads to an overall space requirement, depending on the relatively smaller number of transformers among other factors.

Among other aspects, the storage area depends on:

   i. Material consumption
  ii. Nature of material
 iii. Size and shape of item
  iv. Demand rate
   v. Manufacturing rate
  vi. Adopted manufacturing cycle
 vii. Storage quantity
viii. Inventory holding cost

If optimally occupied space requirement increases, it may lead to an increase in the total related cost. Such increases can be due to the variation in some operational factors. A manufacturing entrepreneur is usually interested in bringing the total relevant cost to an original level in order to become competitive and also to run the business successfully. This objective is also possible/feasible by bringing the space requirement to the original level by focusing on the interaction of various operational factors. Thus specialized discussion is required for manufacturing entrepreneurship along with space consideration.

Manufacturing entrepreneurship also lies in knowing the precise space requirement. Particularly in the context of work-in-process inventory, precise determination of a maximum production-inventory level helps a lot in space requirement planning. This maximum inventory level changes with respect to time depending on a fluctuation in operational factors. Appropriate formulation and analysis would be necessary to determine policies in such a manufacturing scenario.

It is possible to relate the maximum production-inventory level to the space requirement at any stage of manufacturing. For instance, components or items need to be placed between one facility and the next facility. Space requirement may depend on:

a. Production rate of one facility
b. Consumption rate for the next facility
c. Stacking method of items, if any
d. Space needed between rows of items, if any
e. Length and width (or maximum dimension of the item depending on the case) or diameter of the product
f. Significance (if any) of the vertical dimension

In addition to the aforementioned relevant aspects, the manufacturing space requirement between the facilities varies depending on the type of product or industry. However, it is possible to relate the maximum production-inventory level to the stated space requirement for a particular application. Appropriate generalization can also be made concerning the formulation and related analysis for an estimated maximum production-inventory level.

1.2   MANUFACTURING QUALITY

Quality in the context of production or manufacture has many facets, and it depends on the following aspects among others:

i. Nature of product
ii. Processing operation
iii. Production facility
iv. Workmanship

1.2.1   QUALITY CONSIDERATION

Quality can be built into the product for its longer life. However, from an operational or entrepreneurial point of view, a frequent activity involves the rejection during production process. A practical measure can be taken to find out or estimate the proportion of acceptable components or products in a manufacturing lot. Since this proportion varies from company to company, and also from time to time within a manufacturing organization, it may be included in the analysis and effects. Variation of this measure can affect the space requirement also and its suitable arrangement. If such a quantitative information is available to the entrepreneurial manager or engineer, it directly or indirectly helps in practice decisions in an organization.

For example, in sheet metal cutting and forming, a variety of forms can be produced. While doing so, defects may occur and the products/components might be rejected. Occurrence of the defects and their numbers may depend on:

a. Sheet metal composition
b. Equipment used
c. Tool adopted
d. Surface finish of the tool
e. Maintenance of the equipment
f. Properties of tool material
g. Properties of raw material
h. Adjustment of process parameters
i. Worker skills

Because of the occurrence of defects, more components need to be manufactured to meet the demand for acceptable items. This also prolongs the production time and associated costs. In addition to the production time cost/unit production cost, it also has a bearing on rework efforts. Other aspects include the scrap quantity and scrap disposal efforts as well as the related costs.

In a case where the number of defects is more, the proportion of acceptable components/products will be relatively less. On the other hand, this proportion will be greater, if less number of defects occur. Accordingly, the space requirement for keeping the items varies. The manufacturing space requirement depends on the maximum production-inventory level. Sometimes rejected pieces also need to be kept for certain time. This may happen because of:

i. Third-party inspection in near future
ii. Record keeping
iii. Quality analysis
iv. Remedial measures
v. Necessary approval

As mentioned earlier, an appropriate generalization can be made concerning the formulation and related analysis for an estimated maximum production-inventory level. Therefore, there is a need for specific inclusion of the proportion of acceptable components in a manufacturing batch, as shown in Figure 1.6. This may be useful to obtain the revised maximum production-inventory level and subsequent analysis pertaining to the manufacturing entrepreneurship.

1.2.2　Outcomes Concerning Quality

As shown in Figure 1.7, consider an item/component of size: 1×1 m.

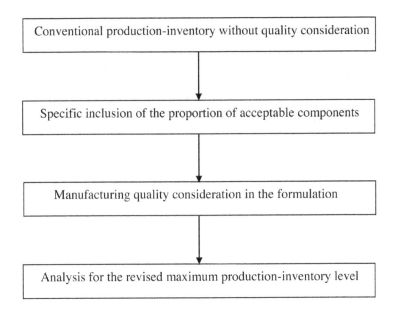

FIGURE 1.6 Approach for manufacturing quality consideration.

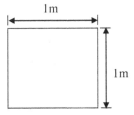

FIGURE 1.7 An item of size 1 × 1 m.

This item/component is produced by a predecessor facility and fed as input for the successor facility as represented by Figure 1.8 along with the intermediate space.

To understand the quality criterion in the present context, assume that the predecessor facility generates some defective components also that are scrapped and not fed to the successor facility. Since the defective items are scrapped, there might not be any need to store them in the intermediate space between the two relevant facilities, i.e., the predecessor facility and the successor facility. However, before the final disposal, such defective pieces may be stored temporarily for relatively short time elsewhere in the organization in case where it is required.

Considering the acceptable items, suppose that the four items are accumulated in the intermediate effective space in certain period. Since the item is of size 1 × 1 m = 1 m², and the available/effective/intermediate space between the two facilities is equivalent to 2 × 2 m = 4 m², the maximum production-inventory of four items can be placed in this space.

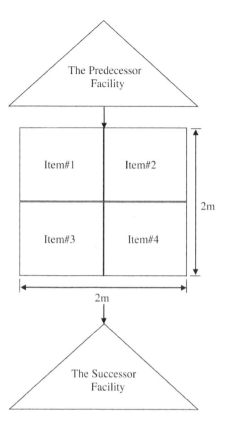

FIGURE 1.8 Available space between the two facilities.

Now, with a sincere effort for quality improvement, the number of acceptable components increases in certain period. If the maximum inventory of five items or more is accumulated, the available intermediate space is insufficient to place them. Therefore, it is of practical and analytical interest to include the quality aspects for the space requirement and the associated planning. The variation in terms of the quality criterion needs to be analyzed in the present manufacturing context.

As shown in Figure 1.9, the manufacturing output is associated with the two possibilities:

  i. Quality improvement
  ii. Quality deterioration

Quality improvement might happen because of:

  a. Deployment of better skilled people
  b. Suitable training of employees
  c. Deployment of better production facilities
  d. Superior maintenance procedures

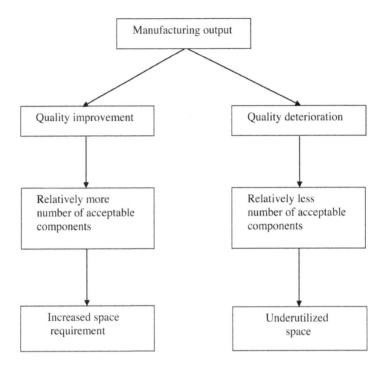

FIGURE 1.9 Possible outcomes concerning quality.

    e. Development of appropriate culture

    f. Appropriate incentives

    g. Overall motivation

Quality deterioration might happen because of:

    i. Improper incentive schemes

    ii. Absence of motivation and appropriate culture

    iii. Old facilities

    iv. Lack of suitable training

    v. Inappropriate maintenance procedures

    vi. Lack of desired skill level

In practical situations, both quality improvement and quality deterioration might be observed in an organization from time to time. Among other reasons, both breakdown and preventive maintenance have an effect on the occurrence of quality scenario in an industrial organization.

In the present context, if quality improves, a number of acceptable components are generated by a facility. On the other hand, if quality deteriorates, the number of defects increases. This increase results into a relatively less number of acceptable components. Because the temporary storage for accepted work-in-process-inventory is desired, there is less requirement of space. This also results into an idle space, and therefore a possible outcome might be in the form of an underutilized space.

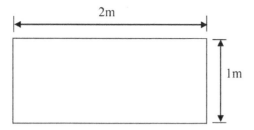

FIGURE 1.10   An item of size 2 × 1 m.

1.2.3   OCCUPIED SPACE

Generally speaking, the space requirement depends on the dimensions/design of the specific component/product. For example, an item can be of rectangular size such as 2 × 1 m, as shown in Figure 1.10.

On comparison with Figure 1.8, instead of four items, only two items can occupy the available/effective space between the two facilities as represented in Figure 1.11.

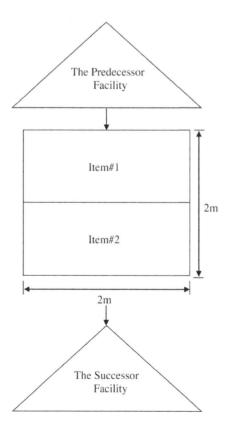

FIGURE 1.11   Available space occupied by the two items.

In the context of space requirement, occupied space varies from case to case; therefore, the maximum production-inventory generated during the process is a suitable parameter for a generalized analysis. Thus, the maximum inventory generated during the manufacturing cycle can be analyzed better in the context of a generalized approach for analysis. For the space requirement and analysis, such a parameter can easily be used for specific component/product related to a specific industry.

1.3 PRODUCTION BACKORDERS

Production or manufacturing backorders might be accumulated because of the scenario such as:

a. Facility is engaged in some other work, and therefore the backlog concerning a particular product accumulates, which also needs to be processed along with the regular work.
b. A specific person is not available, and delay in manufacture might result into backorder or backlog.
c. Non-availability of suitable space for temporary storage of components/products may lead to production backorders.

1.3.1 SUITABLE APPROACH

Figure 1.12 represents an approach for the production backorders inclusion. When manufacturing backorders are accumulated, provision is to be made for their specific inclusion in the formulation and subsequent analysis for the related maximum production-inventory level.

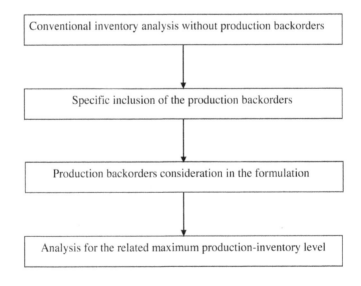

FIGURE 1.12 Approach for the production backorders inclusion.

1.3.2 SPECIFIC INCLUSION

In cases where production backorders are to be specifically included in the formulation, it may be useful for professionals/entrepreneurs in decision making related to such practical situations, where backorders can also be planned well in advance periodically. Such inclusion can also help in the planning of manufacturing space requirement if such problems in a production organization occur. As explained earlier, the space occupied by the components/items depends on the size/design of the specific product of an industry. However, maximum inventory generated during the production process along with backorders can be an appropriate factor for a generalized approach and analysis. This significant aspect has been rigorously included in the present work.

This book is organized in five chapters with the following content:

1. Introduction
   1.1. Manufacturing space
   1.2. Manufacturing quality
   1.3. Production backorders
2. Space consideration
   2.1. Conceptualization
   2.2. Formulation
   2.3. Analysis
3. Quality aspects
   3.1. Basic understanding
   3.2. Mathematical treatment
   3.3. Examples and generalization
4. Backorders
   4.1. Maximum inventory level
   4.2. Entrepreneurial application
   4.3. Quality inclusion
5. Conclusion
   5.1. Useful results
   5.2. Innovation efforts
   5.3. Future scope

This chapter has the introductory description. After discussing commercial and technical aspects of entrepreneurship, it briefly mentions manufacturing and services entrepreneurship. Manufacturing entrepreneurship specifically relates to the production or manufacture of items. Manufacturing space is conceptualized in the context of work-in-process inventory pertaining to the component production. It is discussed with reference to raw material and final product, in addition to the wok-in-progress. Quality and backorders are introduced in the production/manufacturing context. Occupied space is visualized in terms of manufacturing quality consideration. Suitable approach is described for specific inclusion of the production backorders. This chapter especially establishes a need to relate the maximum production-inventory level to the space requirement in order to have a generalized approach.

The second chapter deals with the space consideration. Space is conceptualized with the diameter being a relevant dimension for a cylindrical item in order to keep such items in certain available space. After discussing various assumptions, this chapter visualizes intermediate space for the temporary storage of items including the work-in-progress. Additionally, a manufacturing cycle is depicted along with the relevant parameters. Furthermore, to develop a general approach, a formulation for maximum production-inventory level is needed in the context of batch manufacturing. After providing the relevant formulation, this chapter presents numerous examples and makes rigorous analysis, considering the operational factors or parameters in the context of manufacturing/production entrepreneurship.

The third chapter analyzes in detail quality aspects in manufacture and presents a basic understanding, including the possibilities associated with the manufacturing quality criterion. Mathematical treatment is given considering a manufacturing cycle with quality aspects. Optimal maximum production-inventory level is obtained with an objective of minimizing the total related cost. Its variation with respect to the quality level is discussed. Percentage reduction in maximum inventory level with respect to that in the proportion of acceptable components in a manufacturing batch is tabulated. Similarly, the percentage increase is described. In a business environment, different operational factors/manufacturing parameters change with respect to time. Few examples along with the generalization are provided.

The fourth chapter attempts to understand backorders in production scenario in the context of manufacturing entrepreneurship along with space consideration. Shortfall in the desired demand indicates backorders, and these are incorporated in obtaining the maximum inventory level. The effects of variation in backordering cost on the maximum inventory level are examined. An increase and a reduction in the maximum inventory level are discussed with respect to an annual backordering cost per unit item. Various entrepreneurial applications are discussed considering different factors/manufacturing parameters. Quality inclusion along with backorders in manufacture of a product is also analyzed after the formulation concerning the total related cost.

Finally, the fifth chapter relates to the conclusion and mentions some useful comments/results with reference to the space consideration. Most of the progressive companies are continuously striving for innovation. Innovation efforts in the context of manufacture of such an item are discussed. Furthermore, an approach is analyzed by which the innovative product can be clubbed with the regular product in terms of space consideration, specifically for manufacture. For an assessment of space requirement, production-inventory build-up of two items is described. Additionally, the scenario is discussed when an innovative item is handled independently. Future scope is also discussed highlighting various applications of the suggested concepts in this book.

2 Space Consideration

In this book, space refers to the area available at the outlet side of one facility and before the next facility, and also at various relevant places within the manufacturing organization.

2.1 CONCEPTUALIZATION

To generate the concept of manufacturing space, a related dimension of the product or component should be observed in the context of space required for keeping that item in the location. For example, diameter may be the relevant dimension for a cylindrical item for keeping it in certain available space as shown in Figure 2.1.

2.1.1 ASSUMPTIONS

This representation is on the basis of the following assumptions:

    i. The cylindrical item is kept vertically.
    ii. No space is needed between any two items.
    iii. There is no stacking of items, one over the other.
    iv. There are 4 rows of the 3 items each, thus a maximum of 12 items.
    v. Space refers to the horizontal space.
    vi. Related diameter is 300 mm.

With such assumptions, the space or area required for keeping the items can be found as:

$$\text{Length} = 300 \times 4$$
$$= 1200 \text{ mm}$$
$$= 1.2 \text{ m}$$
$$\text{Width} = 300 \times 3$$
$$= 900 \text{ mm}$$
$$= 0.9 \text{ m}$$
$$\text{Space or area} = \text{Length} \times \text{Width}$$
$$= 1.2 \times 0.9$$
$$= 1.08 \text{ m}^2$$

Such intermediate space might be available between predecessor and successor facility in the manufacturing location, as shown in Figure 2.2.

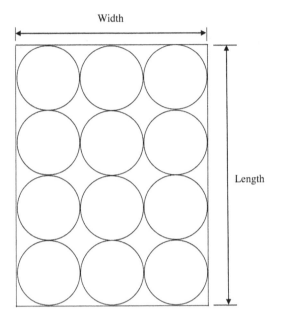

FIGURE 2.1   Representation for occupied space.

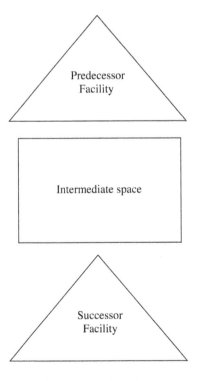

FIGURE 2.2   Intermediate space between two facilities.

2.1.2 Intermediate Space

On the basis of certain policy, the number of components is manufactured by the predecessor facility for subsequent processing on the successor facility. An intermediate space is utilized for keeping certain number of components. As discussed earlier, the space or area, i.e., 1.08 m² corresponds to the 12 items of a particular type. The space requirement differs for another variety or industry type. In the context of a specific case, it is also affected by:

a. Industrial practice of stacking the items, i.e., whether one item over the other is being kept.
b. Need (if any) of space between the two items.

However, to provide a generalization, maximum production-inventory level needs to be obtained. Space requirement can be estimated corresponding to that for a specific component or industry type. A formulation for maximum production-inventory level is needed in the context of batch manufacturing in order to develop a general approach.

2.2 FORMULATION

Figure 2.3 represents the work-in-process since an output of the predecessor facility is continuously going as an input for the successor facility.

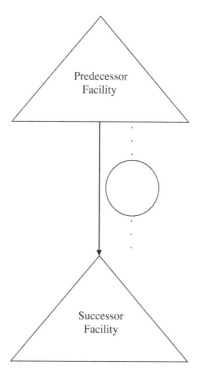

FIGURE 2.3   Work-in-process.

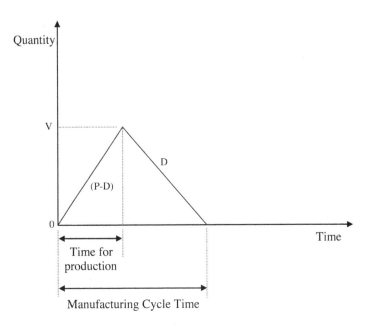

FIGURE 2.4   One manufacturing cycle.

2.2.1   MANUFACTURING CYCLE

One manufacturing cycle is shown in Figure 2.4 with the used notation as follows:

P = Manufacturing rate of the predecessor facility in units per year
D = Demand rate of the successor facility in units per year
V = Maximum production-inventory level

In this context,, the manufacturing rate of predecessor facility is higher than the demand rate of successor facility. In cases where it is not so, there is no question of inventory accumulation at the intermediate space, and such a space might not be needed.

During the time for production or manufacture in this cycle, both the manufacturing rate (of predecessor facility) and demand rate (of successor facility) are active. Therefore, the work-in-process inventory build-up rate is (P – D). This continues until the maximum production-inventory level V is attained in this cycle. Afterwards (i.e., after the time for manufacture in a cycle is over), only the demand rate of the successor facility is active. Therefore the maximum inventory level V starts decreasing at the rate D. The cycle completes when this intermediate work-in-process inventory becomes zero. After this one manufacturing cycle is over, the similar cycle starts.

Now:

Manufacturing cycle time = Time for manufacture + Time for consumption
of maximum inventory level

Or Cycle time is,

$$= \frac{V}{(P-D)} + \frac{V}{D}$$

$$= \frac{VD + V(P-D)}{D(P-D)}$$

$$= \frac{VP}{D(P-D)}$$

$$= \frac{V}{D(1-D/P)}$$

Now:

Number of cycles in a year $= 1/(\text{cycle time})$

Or: Number of cycles in a year,

$$= \frac{D(1-D/P)}{V}$$

Since facility setup cost is incurred for each cycle, an annual facility setup cost is evaluated as the multiplication of such setup cost and the number of cycles in a year. Therefore, annual facility setup cost,

$$= \frac{CD(1-D/P)}{V} \tag{2.1}$$

where C = Facility setup cost.

As the average inventory is,

$$\frac{V}{2}$$

Annual production – inventory carrying cost,

$$= \frac{VI}{2} \tag{2.2}$$

where I = Annual production-inventory carrying cost per unit.

Adding the expressions (2.2) and (2.1), the total related cost (E) is formulated as:

$$E = \frac{VI}{2} + \frac{CD(1-D/P)}{V} \tag{2.3}$$

With an objective of minimizing the total related cost, differentiating with respect to V and equating to zero,

$$\frac{I}{2} - \frac{CD(1-D/P)}{V^2} = 0$$

Or:

$$\frac{CD(1-D/P)}{V^2} = \frac{I}{2}$$

Or:

$$V^2 = \frac{2CD(1-D/P)}{I}$$

Or:

$$V^* = \sqrt{\frac{2CD(1-D/P)}{I}} \tag{2.4}$$

The above expression will give an optimal maximum production-inventory level while implementing the minimum total related cost.

From Eq. (2.4):

$$CD(1-D/P) = \frac{IV^{*2}}{2}$$

Substituting in Eq. (2.3), the total related cost in the present context can be derived as:

$$E^* = \frac{IV^*}{2} + \frac{IV^*}{2}$$

Or:

$$E^* = IV^* \tag{2.5}$$

While implementing this total related cost, suitable space is required for temporary storage of optimal maximum production-inventory level. The temporary storage of items might be required for the stages such as:

a. Manufacturing
b. Inspection
c. Testing
d. Packaging
e. Outbound logistics

With appropriate perception of operational factors, this approach can also be utilized for interfaces such as:

i. Arrival of input items and release to the first facility for starting the production and subsequent processes
ii. Last facility and subsequent storage
iii. Manufacturing facility and subsequent inspection or testing stage
iv. Packaging and outbound logistics

2.2.2 REPRESENTATION FOR TEMPORARY STORAGE

Representation for temporary storage is illustrated with the following example.

Example 2.1

Consider the following information:

Manufacturing rate of the predecessor facility in units per year, P = 1250
Demand rate of the successor facility in units per year, D = 750
Annual production-inventory carrying cost per unit, I = ₹50
Facility setup cost, C = ₹75

Now:
From the Eq. (2.4), optimal maximum production-inventory level can be obtained as:

$$V^* = \sqrt{\frac{2CD(1 - D/P)}{I}}$$

$$= \sqrt{\frac{2 \times 75 \times 750 \times (1 - 750/1250)}{50}}$$

$$= 30 \text{ units}$$

From the Eq. (2.5), optimal total related cost is obtained as follows:

$$E^* = IV^*$$

$$= 50 \times 30$$

$$= ₹1500$$

The information of maximum level of inventory of 30 units helps in the space requirement planning. There can be multiple ways of keeping these units depending on the product design and dimensions. However, considering the previous discussion (i.e., the cylindrical item kept vertically with the related diameter 300 mm), one way of keeping the 30 items is shown in Figure 2.5.

Width

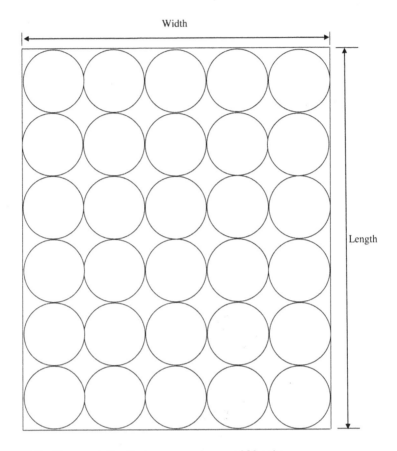

Length

FIGURE 2.5   Representation for temporary storage of 30 units.

The space or area required for keeping the items can be found as:

$$Length = 300 \times 6$$
$$= 1800 \text{ mm}$$
$$= 1.8 \text{ m}$$
$$Width = 300 \times 5$$
$$= 1500 \text{ mm}$$
$$= 1.5 \text{ m}$$
$$Area = Length \times Width$$
$$= 1.8 \times 1.5$$
$$= 2.70 \text{ m}^2$$

The requirement of space depends on the type of industry or product, i.e., the product dimensions among other aspects. However, the maximum production-inventory level (V) is the fundamental factor. On the basis of the V value, the space requirement can be estimated or visualized for specific cases of manufacturing industry. Therefore, the subsequent discussion/analysis mainly focuses on this derived parameter V.

2.3 ANALYSIS

A maximum production-inventory level depends on the following factors:

a. Facility setup cost
b. Inventory carrying cost
c. Facility production rate
d. Demand rate for the subsequent use

These factors vary with respect to time or the specific need. Facility setup cost has a bearing on setup time. Setup time may increase because of the workers' inexperience among other issues. On the other hand, setup time reduces because of reasons such as an increased skill level of the workers. Upward as well as downward fluctuation in the inventory carrying cost can happen because of an opportunity cost and/or storage cost variation among other factors. Variation in the overall facility production rate might happen because of human resources in certain case. In the event of packaging activity, for instance, the effective demand rate might fluctuate owing to the availability of packaging material (or corresponding space to keep it temporarily) or human resources among other practical issues in the manufacturing concern.

Since variation in these factors occur in the real environment, an idea concerning their effect on the space requirement, helps a lot in the context of manufacturing entrepreneurship.

Example 2.2

With the following information:

Manufacturing rate, $P = 1250$
Demand rate, $D = 750$
Annual inventory carrying cost per unit, $I = ₹50$
Facility setup cost, $C = ₹75$

Value of maximum inventory level, V is obtained as equivalent to 30 units with the use of Eq. (2.4).

Now, if the facility setup cost increases by 20%, then the increased setup cost is as given below:

$$C_1 = 75 \times 1.2 = ₹90$$

And the maximum inventory level (V) is found as:

$$\sqrt{\frac{2C_1D(1-D/P)}{I}}$$

$$= \sqrt{\frac{2 \times 90 \times 750 \times (1 - 750/1250)}{50}}$$

$$= 32.86 \text{ units}$$

$$\approx 33 \text{ units}$$

That is approximately 10% increase in the V value.
    For a general approach:
    Increase in the maximum inventory level,

$$V_1 - V = \sqrt{\frac{2C_1D(1-D/P)}{I}} - \sqrt{\frac{2CD(1-D/P)}{I}}$$

where:
    C_1 = Increased setup cost
    V_1 = Corresponding increased maximum inventory level

Now:

$$V_1 - V = \sqrt{\frac{2CD(1-D/P)}{I}}\left[\sqrt{\frac{C_1}{C}} - 1\right]$$

Percentage increase in V,

$$= \frac{(V_1 - V)}{V} * 100$$

$$= \left[\sqrt{\frac{C_1}{C}} - 1\right] * 100$$

Substituting the value of C_1 as:

$$C_1 = C\left(1 + \frac{W}{100}\right)$$

where W = % increase in C.

Now:
    % increase in V,

$$= \left[\sqrt{\left(1 + \frac{W}{100}\right)} - 1\right] * 100$$

TABLE 2.1

% Increase in V with respect to % Increase in C

S. No.	W	$\left[\sqrt{\left(1+\dfrac{W}{100}\right)}-1\right]*100$
1	10	4.88
2	20	9.54
3	30	14.02
4	40	18.32
5	50	22.47

Table 2.1 shows the % increase in V with respect to the % increase in C. The % increase in V is lower than half of the corresponding W values.

If a facility setup cost reduces, the reduction in the maximum inventory level,

$$V - V_1 = \sqrt{\frac{2CD(1 - D/P)}{I}} - \sqrt{\frac{2C_1 D(1 - D/P)}{I}}$$

where:

C_1 = Reduced setup cost

V_1 = Corresponding reduced maximum inventory level

Now:

$$V - V_1 = \sqrt{\frac{2CD(1 - D/P)}{I}}\left[1 - \sqrt{\frac{C_1}{C}}\right]$$

Percentage reduction in V,

$$= \frac{(V - V_1)}{V} * 100$$

$$= \left[1 - \sqrt{\frac{C_1}{C}}\right] * 100$$

Substituting the value of C_1 as:

$$C_1 = C\left(1 - \frac{W}{100}\right)$$

where W = % reduction in C.

Now:

% reduction in V,

$$= \left[1 - \sqrt{\left(1 - \frac{W}{100}\right)}\right] * 100$$

TABLE 2.2

% Reduction in V with respect to % Reduction in C

S. No.	W	$\left[1-\sqrt{\left(1-\dfrac{W}{100}\right)}\right] * 100$
1	10	5.13
2	20	10.56
3	30	16.33
4	40	22.54
5	50	29.29

Table 2.2 shows the % reduction in V with respect to the % reduction in C. The % reduction in V is more than half of the corresponding W values.

On comparison with Table 2.1, the percentage variation in V is relatively higher. This can also be verified as follows:

$$\left[1-\sqrt{\left(1-\frac{W}{100}\right)}\right] * 100 > \left[\sqrt{\left(1+\frac{W}{100}\right)} - 1\right] * 100$$

Or:

$$1-\sqrt{\left(1-\frac{W}{100}\right)} > \sqrt{\left(1+\frac{W}{100}\right)} - 1$$

Or:

$$2 > \sqrt{\left(1+\frac{W}{100}\right)} + \sqrt{\left(1-\frac{W}{100}\right)}$$

Or:

$$2^2 > \left[\sqrt{\left(1+\frac{W}{100}\right)} + \sqrt{\left(1-\frac{W}{100}\right)}\right]^2$$

Or:

$$4 > 1 + \frac{W}{100} + 1 - \frac{W}{100} + 2\sqrt{1-\left(\frac{W}{100}\right)^2}$$

Or:

$$4 > 2 + 2\sqrt{1-\left(\frac{W}{100}\right)^2}$$

Or:

$$2 > 2\sqrt{1 - \left(\frac{W}{100}\right)^2}$$

Or:

$$1 > \sqrt{1 - \left(\frac{W}{100}\right)^2}$$

For all practical values of W (i.e., less than 100), the above expression can be verified. For instance:

$$W = 15:$$

$$\sqrt{1 - \left(\frac{W}{100}\right)^2} = 0.989$$

$$W = 85:$$

$$\sqrt{1 - \left(\frac{W}{100}\right)^2} = 0.527$$

Example 2.3

The given data are as follows:

Manufacturing rate, $P = 1250$
Demand rate, $D = 750$
Annual inventory carrying cost per unit, $I = ₹50$
Facility setup cost, $C = ₹75$

Value of maximum inventory level, V, is obtained as equivalent to 30 units with the use of Eq. (2.4).

Now, if there is an increase in annual inventory carrying cost per unit by 20%, the increased value is given as:

$$I_1 = 50 \times 1.2 = ₹60$$

And the maximum inventory level (V) is found as:

$$\sqrt{\frac{2CD(1 - D/P)}{I_1}}$$

$$= \sqrt{\frac{2 \times 75 \times 750 \times (1 - 750/1250)}{60}}$$

$$= 27.39 \text{ units}$$

That is approximately 9% reduction in the V value.

For a general approach:

Reduction in the maximum inventory level,

$$V - V_1 = \sqrt{\frac{2CD(1-D/P)}{I}} - \sqrt{\frac{2CD(1-D/P)}{I_1}}$$

where:

I_1 = Increased annual inventory carrying cost per unit

V_1 = Corresponding reduced maximum inventory level

Now:

$$V - V_1 = \sqrt{\frac{2CD(1-D/P)}{I}}\left[1 - \sqrt{\frac{I}{I_1}}\right]$$

Percentage reduction in V,

$$= \frac{(V - V_1)}{V} * 100$$

$$= \left[1 - \sqrt{\frac{I}{I_1}}\right] * 100$$

Substituting the value of I_1 as:

$$I_1 = I\left(1 + \frac{W}{100}\right)$$

where W = % increase in I.

Now:

% reduction in V,

$$= \left[1 - \sqrt{\frac{1}{\left(1 + \frac{W}{100}\right)}}\right] * 100$$

Table 2.3 shows the % reduction in V with respect to the % increase in I. The % reduction in V is lower than half of the corresponding W values.

If an annual inventory carrying cost per unit reduces, the increase in the maximum inventory level,

$$V_1 - V = \sqrt{\frac{2CD(1-D/P)}{I_1}} - \sqrt{\frac{2CD(1-D/P)}{I}}$$

where:

I_1 = Reduced annual inventory carrying cost per unit

V_1 = Corresponding increased maximum inventory level

TABLE 2.3
% Reduction in V with respect to % Increase in I

S. No.	W	$\left[1-\sqrt{\dfrac{1}{\left(1+\dfrac{W}{100}\right)}}\right]*100$
1	15	6.75
2	25	10.56
3	35	13.93
4	45	16.95
5	55	19.68

Now:

$$V_1 - V = \sqrt{\frac{2CD(1 - D/P)}{I}}\left[\sqrt{\frac{I}{I_1}} - 1\right]$$

Percentage increase in V,

$$= \frac{(V_1 - V)}{V} * 100$$

$$= \left[\sqrt{\frac{I}{I_1}} - 1\right] * 100$$

Substituting the value of I_1 as:

$$I_1 = I\left(1 - \frac{W}{100}\right)$$

where W = % reduction in I.

Now:
   % increase in V,

$$= \left[\sqrt{\frac{1}{\left(1 - \dfrac{W}{100}\right)}} - 1\right] * 100$$

Table 2.4 shows the % increase in V with respect to the % reduction in I. The % reduction in V is more than half of the corresponding W values.

In comparison with Table 2.3, the percentage variation in V is higher.

TABLE 2.4

% Increase in V with respect to % Reduction in I

S. No.	W	$\left[\sqrt{\dfrac{1}{\left(1-\dfrac{W}{100}\right)}}-1\right]*100$
1	15	8.47
2	25	15.47
3	35	24.03
4	45	34.84
5	55	49.07

Example 2.4

With the following information:

Manufacturing rate, $P = 1250$;
Demand rate, $D = 750$;
Annual inventory carrying cost per unit, $I = ₹50$;
Facility setup cost, $C = ₹75$;

Value of maximum inventory level, V is obtained as equivalent to 30 units with the use of Eq. (2.4).

Now, if there is an increase in the manufacturing rate by 20%, the increased value is given as:

$$P_1 = 1250 \times 1.2 = 1500$$

And the maximum inventory level (V) is found as:

$$\sqrt{\frac{2CD(1-D/P_1)}{I}}$$

$$= \sqrt{\frac{2 \times 75 \times 750 \times (1-750/1500)}{50}}$$

$$= 33.54 \text{ units}$$

That is approximately 12% increase in the V value.

For a general approach:
Increase in the maximum inventory level,

$$V_1 - V = \sqrt{\frac{2CD(1-D/P_1)}{I}} - \sqrt{\frac{2CD(1-D/P)}{I}}$$

where:

    P_1 = Increased manufacturing rate

    V_1 = Corresponding increased maximum inventory level

Now:

$$V_1 - V = \sqrt{\frac{2CD(1-D/P)}{I}}\left[\sqrt{\frac{(1-D/P_1)}{(1-D/P)}}-1\right]$$

Percentage increase in V,

$$= \frac{(V_1-V)}{V}*100$$

$$= \left[\sqrt{\frac{(1-D/P_1)}{(1-D/P)}}-1\right]*100$$

Substituting the value of P_1 as:

$$P_1 = P\left(1+\frac{W}{100}\right)$$

where W = % increase in P.

Now:

    % increase in V,

$$= \left[\sqrt{\frac{1-D/\{P(1+W/100)\}}{(1-D/P)}}-1\right]*100$$

Table 2.5 shows the % increase in V with respect to the % increase in P. The % increase in V is much less than the corresponding W values for the given P and D.

    In case where the manufacturing rate reduces, the reduction in the maximum inventory level,

$$V - V_1 = \sqrt{\frac{2CD(1-D/P)}{I}} - \sqrt{\frac{2CD(1-D/P_1)}{I}}$$

TABLE 2.5
% Increase in V with respect to % Increase in P

S. No.	W	$\left[\sqrt{\dfrac{1-D/\{P(1+W/100)\}}{(1-D/P)}}-1\right]*100$
1	15	9.35
2	25	14.02
3	35	17.85
4	45	21.06
5	55	23.78

where:

P_1 = Reduced manufacturing rate
V_1 = Corresponding reduced maximum inventory level

Now:

$$V - V_1 = \sqrt{\frac{2CD(1 - D/P)}{I}}\left[1 - \sqrt{\frac{(1 - D/P_1)}{(1 - D/P)}}\right]$$

Percentage reduction in V,

$$= \frac{(V - V_1)}{V} * 100$$

$$= \left[1 - \sqrt{\frac{(1 - D/P_1)}{(1 - D/P)}}\right] * 100$$

Substituting the value of P_1 as:

$$P_1 = P\left(1 - \frac{W}{100}\right)$$

where W = % reduction in P.

Now:

% reduction in V,

$$= \left[1 - \sqrt{\frac{1 - D/\{P(1 - W/100)\}}{(1 - D/P)}}\right] * 100$$

Table 2.6 shows the % reduction in V with respect to the % reduction in P, for the given P and D.

TABLE 2.6

% Reduction in V with respect to % Reduction in P

S. No.	W	$\left[1 - \sqrt{\dfrac{1 - D/\{P(1 - W/100)\}}{(1 - D/P)}}\right] * 100$
1	15	14.25
2	20	20.94
3	25	29.29
4	30	40.24
5	35	56.15

Example 2.5

Using the data as follows:

> Manufacturing rate, $P = 1250$
> Demand rate, $D = 750$
> Annual inventory carrying cost per unit, $I = ₹50$
> Facility setup cost, $C = ₹75$

Value of maximum inventory level, V is obtained as equivalent to 30 units with the use of Eq. (2.4).

Now, if there is an increase in demand rate by 20%, the increased value is as given below:

$$D_1 = 750 \times 1.2 = 900$$

And the maximum inventory level (V) is found as:

$$\sqrt{\frac{2CD_1(1 - D_1 / P)}{I}}$$

$$= \sqrt{\frac{2 \times 75 \times 900 \times (1 - 900 / 1250)}{50}}$$

$$= 27.50 \text{ units}$$

That is approximately 8% reduction in the V value.
For a general approach:
Reduction in the maximum inventory level,

$$V - V_1 = \sqrt{\frac{2CD(1 - D / P)}{I}} - \sqrt{\frac{2CD_1(1 - D_1 / P)}{I}}$$

where:
D_1 = Increased demand rate
V_1 = Corresponding reduced maximum inventory level

Now:

$$V - V_1 = \sqrt{\frac{2CD(1 - D / P)}{I}} \left[1 - \sqrt{\frac{D_1(1 - D_1 / P)}{D(1 - D / P)}} \right]$$

Percentage reduction in V,

$$= \frac{(V - V_1)}{V} * 100$$

$$= \left[1 - \sqrt{\frac{D_1(1 - D_1 / P)}{D(1 - D / P)}} \right] * 100$$

TABLE 2.7

% Reduction in V with respect to % Increase in D

S. No.	W	$\left[1-\sqrt{\dfrac{(1+W/100)\{1-D(1+W/100)/P\}}{(1-D/P)}}\right]*100$
1	15	5.59
2	20	8.35
3	25	11.61
4	30	15.44
5	35	19.92

Substituting the value of D_1 as:

$$D_1 = D\left(1+\frac{W}{100}\right)$$

where W = % increase in D.

Now:

% reduction in V,

$$=\left[1-\sqrt{\frac{(1+W/100)\{1-D(1+W/100)/P\}}{(1-D/P)}}\right]*100$$

Table 2.7 shows the % reduction in V with respect to the % increase in D. The % reduction in V is much less than the corresponding W values for the given P and D.

If the demand rate reduces, the increase in the maximum inventory level,

$$V_1-V = \sqrt{\frac{2CD_1(1-D_1/P)}{I}} - \sqrt{\frac{2CD(1-D/P)}{I}}$$

where:

D_1 = Reduced demand rate
V_1 = Corresponding increased maximum inventory level

Now:

$$V_1-V = \sqrt{\frac{2CD(1-D/P)}{I}}\left[\sqrt{\frac{D_1(1-D_1/P)}{D(1-D/P)}} -1\right]$$

Percentage increase in V,

$$=\frac{(V_1-V)}{V}*100$$

$$=\left[\sqrt{\frac{D_1(1-D_1/P)}{D(1-D/P)}} -1\right]*100$$

TABLE 2.8

% Increase in V with respect to % Reduction in D

S. No.	W	$\left[\sqrt{\dfrac{(1-W/100)\{1-D(1-W/100)/P\}}{(1-D/P)}} - 1\right] * 100$
1	15	2.04
2	20	1.98
3	25	1.55

Substituting the value of D_1 as:

$$D_1 = D\left(1 - \frac{W}{100}\right)$$

where W = % reduction in D.

Now:
% increase in V,

$$= \left[\sqrt{\frac{(1-W/100)\{1-D(1-W/100)/P\}}{(1-D/P)}} - 1\right] * 100$$

Table 2.8 shows the % increase in V with respect to the % reduction in D, for the given P and D.

A considerable success can be achieved when an entrepreneur or entrepreneurial manager handles the fluctuation in operational features well with respect to time. Operational aspects such as demand/production rate, facility setup cost, and an inventory carrying cost may vary. Because of this, the space requirement also varies. If the maximum production-inventory level (V) reduces, there is no need for additional arrangement for space. However, if the value of V increases, there is a requirement for additional space. This might not be easier in many cases. However, if the V value can be brought to the original level by way of altering another feasible operational feature, a considerable success can be achieved in terms of manufacturing entrepreneurship. The V value may increase because of:

i. Demand rate reduction
ii. Facility setup cost increase
iii. Inventory carrying cost reduction
iv. Production rate increase

Example 2.6

With the information as follows:

Manufacturing rate, P = 1250
Demand rate, D = 750
Annual inventory carrying cost per unit, I = ₹50
Facility setup cost, C = ₹75

Value of maximum inventory level, V is obtained as equivalent to 30 units with the use of Eq. (2.4).

Now, if there is the demand rate reduction by 20%, the D value is given as:

$$D_1 = 750 \times 0.8 = 600$$

And the maximum inventory level (V) is found as:

$$\sqrt{\frac{2CD_1(1-D_1/P)}{I}}$$

$$= \sqrt{\frac{2 \times 75 \times 600 \times (1-600/1250)}{50}}$$

$$= 30.59 \text{ units}$$

To bring the V value back to the original level, the following are suitable options:

a. Facility setup cost reduction
b. Facility production rate reduction

Considering the first option, the revised C value can be obtained as:

$$\sqrt{\frac{2 \times C_1 \times 600 \times (1-600/1250)}{50}} = 30$$

Or $C_1 = ₹72.12$

For a general approach:

$$\sqrt{\frac{2C_1D_1(1-D_1/P)}{I}} = \sqrt{\frac{2CD(1-D/P)}{I}}$$

Or:

$$C_1D_1(1-D_1/P) = CD(1-D/P)$$

Substituting the values as:

$$D_1 = D\left(1 - \frac{W}{100}\right)$$

$$C_1 = C\left(1 - \frac{Z}{100}\right)$$

where:
  W = % reduction in D
  Z = % reduction in C

Now:

$$(1-Z/100)(1-W/100)[1-(D/P)(1-W/100)] = (1-D/P)$$

Or:

$$1 - \frac{Z}{100} = \frac{(1 - D/P)}{(1 - W/100)[1 - (D/P)(1 - W/100)]}$$

Or:

$$\frac{Z}{100} = 1 - \frac{(1 - D/P)}{(1 - W/100)[1 - (D/P)(1 - W/100)]}$$

Or:

$$Z = 100\left[1 - \frac{(1 - D/P)}{(1 - W/100)\{1 - (D/P)(1 - W/100)\}}\right]$$

In the discussed example:
  W = 20
  D = 750
  P = 1250

Now Z can be obtained as equivalent to 3.846%, and therefore:

$$C_1 = C\left(1 - \frac{Z}{100}\right) = 75\left(1 - \frac{3.846}{100}\right)$$

Or C_1 = ₹72.12.

Another suitable option can be a reduction in the rate of manufacture. Now:

$$\sqrt{\frac{2CD_1(1 - D_1/P_1)}{I}} = \sqrt{\frac{2CD(1 - D/P)}{I}}$$

Or:

$$(1 - D_1/P_1) = \frac{(1 - D/P)}{(1 - W/100)}$$

Or:

$$\frac{D_1}{P_1} = 1 - \frac{(1 - D/P)}{(1 - W/100)}$$

Or:

$$\frac{D_1}{P_1} = \frac{(D/P) - (W/100)}{(1 - W/100)}$$

Or:

$$\frac{P_1}{D_1} = \frac{(1 - W/100)}{(D/P) - (W/100)}$$

Substituting the values as:

$$D_1 = D\left(1 - \frac{W}{100}\right)$$

$$P_1 = P\left(1 - \frac{Z}{100}\right)$$

where:
W = % reduction in D
Z = % reduction in P
Or:

$$\frac{(1 - Z/100)}{(D/P)(1 - W/100)} = \frac{(1 - W/100)}{(D/P) - (W/100)}$$

Or:

$$1 - \frac{Z}{100} = \frac{(D/P)(1 - W/100)^2}{(D/P) - (W/100)}$$

Or:

$$\frac{Z}{100} = 1 - \frac{(D/P)(1 - W/100)^2}{(D/P) - (W/100)}$$

Or:

$$Z = 100\left[1 - \frac{(D/P)(1 - W/100)^2}{(D/P) - (W/100)}\right]$$

In the discussed example:
W = 20
D = 750
P = 1250

Now Z can be obtained as equivalent to 4%, and therefore:

$$P_1 = P\left(1 - \frac{Z}{100}\right) = 1250\left(1 - \frac{4}{100}\right)$$

Or:

$$P_1 = 1200$$

While comparing with the previous option, the Z value is relatively higher. Such information is useful in selecting the appropriate option for an organization, because the efforts needed (and/or the feasibility) to alter the specific operational factor vary from organization to organization.

Example 2.7

Because of an annual inventory carrying cost per unit (I) reduction, the space requirement increases. An option can be the facility setup cost reduction for similar space.

Now:

$$\sqrt{\frac{2C_1D(1-D/P)}{I_1}} = \sqrt{\frac{2CD(1-D/P)}{I}}$$

Or:

$$\frac{C_1}{C} = \frac{I_1}{I}$$

Or:

$$Z = W$$

where

$$I_1 = I\left(1 - \frac{W}{100}\right)$$

$$C_1 = C\left(1 - \frac{Z}{100}\right)$$

where:
 W = % reduction in I
 Z = % reduction in C

Another option can be the manufacturing rate reduction. That is:

$$\sqrt{\frac{2CD(1-D/P_1)}{I_1}} = \sqrt{\frac{2CD(1-D/P)}{I}}$$

Or:

$$1 - \frac{D}{P_1} = \frac{I_1(1-D/P)}{I}$$

Or:

$$1 - \frac{D}{P_1} = (1 - W/100)(1 - D/P)$$

Or:

$$\frac{D}{P_1} = 1 - (1 - W/100)(1 - D/P)$$

Or:

$$P_1 = \frac{D}{1 - (1 - W/100)(1 - D/P)}$$

Substituting P_1 as:

$$P_1 = P\left(1 - \frac{Z}{100}\right)$$

$$1 - \frac{Z}{100} = \frac{(D/P)}{1 - (1 - W/100)(1 - D/P)}$$

Or:

$$Z = 100\left[1 - \frac{(D/P)}{1 - (1 - W/100)(1 - D/P)}\right]$$

With the information as follows:

Manufacturing rate, P = 1250
Demand rate, D = 750
Annual inventory carrying cost per unit, I = ₹50
Facility setup cost, C = ₹75

And, for W = 10, Z value can be obtained as 6.25, i.e., relatively lower than the previous option because Z = W for that option.

Example 2.8

Because of the increase in facility setup cost, the additional requirement for space arises. Manufacturing rate reduction can be an option to bring back the V value to an original level.
Now:

$$C_1 = C\left(1 + \frac{W}{100}\right)$$

$$P_1 = P\left(1 - \frac{Z}{100}\right)$$

where:
W = % increase in C
Z = % reduction in P

And:

$$\sqrt{\frac{2C_1 D(1 - D/P_1)}{I}} = \sqrt{\frac{2CD(1 - D/P)}{I}}$$

Or:

$$C_1(1 - D/P_1) = C(1 - D/P)$$

Or:

$$1 - \frac{D}{P_1} = \frac{(1 - D/P)}{(1 + W/100)}$$

Or:

$$\frac{D}{P_1} = 1 - \frac{(1-D/P)}{(1+W/100)}$$

Or:

$$\frac{D}{P_1} = \frac{(W/100)+(D/P)}{(1+W/100)}$$

Or:

$$P_1 = \frac{D(1+W/100)}{(W/100)+(D/P)}$$

Or:

$$1 - \frac{Z}{100} = \frac{(D/P)(1+W/100)}{(W/100)+(D/P)}$$

Or:

$$Z = 100\left[1 - \frac{(D/P)(1+W/100)}{(W/100)+(D/P)}\right]$$

Considering the previous Example, the Z value can be obtained as 10 for W = 20.

Example 2.9

The V value increases because of the upward variation in the rate of manufacture. Reduction in facility setup cost can be a response for the objective of maintaining a similar maximum production-inventory level.
Now:

$$P_1 = P\left(1 + \frac{W}{100}\right)$$

$$C_1 = C\left(1 - \frac{Z}{100}\right)$$

where:
   W = % increase in P
   Z = % reduction in C

And:

$$\sqrt{\frac{2C_1 D(1-D/P_1)}{I}} = \sqrt{\frac{2CD(1-D/P)}{I}}$$

Or:

$$C_1(1-D/P_1) = C(1-D/P)$$

Or:

$$1 - \frac{Z}{100} = \frac{(1 - D/P)}{(1 - D/P_1)}$$

Or:

$$Z = 100 \left[1 - \frac{(1 - D/P)}{(1 - D/P_1)} \right]$$

Or:

$$Z = 100 \left[\frac{(D/P) - (D/P_1)}{(1 - D/P_1)} \right]$$

Or:

$$Z = 100 \left[\frac{D(P_1/P) - D}{P_1 - D} \right]$$

Or:

$$Z = 100 \left[\frac{D(1 + W/100) - D}{P(1 + W/100) - D} \right]$$

Or:

$$Z = \frac{DW}{P(1 + W/100) - D}$$

With the use of the following values:

   Manufacturing rate, P = 1250
   Demand rate, D = 750

And for W = 15, the Z value can be obtained as equivalent to 16.36.

   In the context of manufacturing entrepreneurship with specific reference to the space consideration, these few examples show how one can deal with the scenario when the V value increases because of:

   i. Demand rate reduction
   ii. Inventory carrying cost reduction
   iii. Facility setup cost increase
   iv. Manufacturing rate increase

Few related options are explained that concern with the possible/feasible variation in the operational factors.

3 Quality Aspects

Quality has many facets and can be built-in during the stages such as:

a. Research and development phase of the proposed component/product
b. Detailed design of various components/product
c. Selection of appropriate material/initial processing
d. Substitution of existing material for enhanced quality
e. Selection of appropriate technology for manufacturing
f. Choice and installation of suitable production facilities

However, presently, the focus is on the quality scenario when actual production is going on a regular basis in a manufacturing organization. A practical and convenient measurement in such cases is the proportion of acceptable components in a certain production lot. This has been used in order to provide a basic understanding and also the mathematical treatment of this topic.

3.1 BASIC UNDERSTANDING

Consider a manufacturing facility as shown in Figure 3.1 with the three possibilities.

In the most ideal scenario, there are no defective components produced by a manufacturing facility; i.e., all are acceptable components. However, this might not happen very often. Other possibilities may be grouped together; i.e., some defective components are also produced, either reworked or rejected. In the case of rework, the component might pass through the facility again. In the case of rejected component, the facility is used again to compensate for shortfall.

For example, assume that 900 components are required in a lot and the proportion of acceptable components in a lot is 0.9. Now, the acceptable components are:

$$900 \times 0.9 = 810$$

And the requirement of 900 is not fulfilled. In order to achieve this, there is a need to produce at least 1000 components so that:

$$1000 \times 0.9 = 900$$

Alternatively, the effective production or manufacturing rate corresponding to the demand of acceptable components would be:

$$yP$$

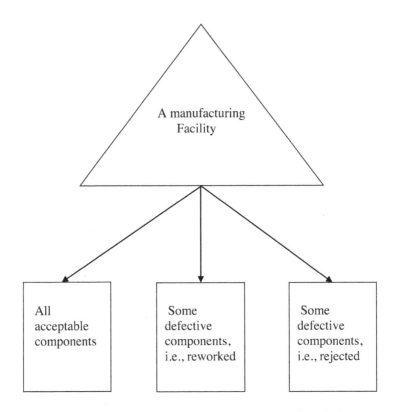

FIGURE 3.1 Possibilities associated with the manufacturing quality criterion.

where:
    y = Proportion of acceptable components in a manufacturing batch
    P = Manufacturing rate of the facility in units per year

In a batch manufacture:

$$yP > D$$

where D = Demand rate of the acceptable components in units per year.

3.1.1 MANUFACTURING CYCLE WITH QUALITY CRITERION

One manufacturing cycle is shown in Figure 3.2 with the additional notation as follows:

    V = Maximum production – inventory level concerning acceptable components

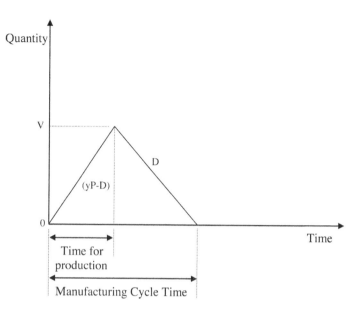

FIGURE 3.2   One manufacturing cycle with quality aspects.

3.2   MATHEMATICAL TREATMENT

With reference to Figure 3.2,
    Cycle time,

$$= \frac{V}{(yP - D)} + \frac{V}{D}$$

$$= \frac{VD + V(yP - D)}{D(yP - D)}$$

$$= \frac{VyP}{D(yP - D)}$$

$$= \frac{V}{D(1 - D/yP)}$$

Now:

$$\text{Number of cycles in a year} = 1/(\text{Cycle time})$$

Or:

    Number of cycles in a year,

$$= \frac{D(1 - D/yP)}{V}$$

Since facility setup cost is incurred for each cycle, an annual facility setup cost is evaluated as the multiplication of such setup cost and the number of cycles in a year. Therefore:

Annual facility setup cost,

$$= \frac{CD(1 - D/yP)}{V} \tag{3.1}$$

where C = Facility setup cost.

As the average inventory is,

$$\frac{V}{2}$$

Annual production-inventory carrying cost,

$$= \frac{VI}{2} \tag{3.2}$$

where I = Annual production-inventory carrying cost per unit.

Adding the expression (3.2) and (3.1), the total related cost (E) is formulated as:

$$E = \frac{VI}{2} + \frac{CD(1 - D/yP)}{V} \tag{3.3}$$

With an objective of minimizing the total related cost, differentiating with respect to V and equating to zero,

$$\frac{I}{2} - \frac{CD(1 - D/yP)}{V^2} = 0$$

Or:

$$\frac{CD(1 - D/yP)}{V^2} = \frac{I}{2}$$

Or:

$$V^2 = \frac{2CD(1 - D/yP)}{I}$$

Or:

$$V* = \sqrt{\frac{2CD(1 - D/yP)}{I}} \tag{3.4}$$

The above expression will give an optimal maximum production-inventory level while implementing the minimum total related cost.

From Eq. (3.4):

$$CD(1 - D/yP) = \frac{IV*^2}{2}$$

Substituting in Eq. (3.3), the total related cost in the present context can be derived as:

$$E* = \frac{IV*}{2} + \frac{IV*}{2}$$

Or:

$$E* = IV* \qquad\qquad (3.5)$$

Example 3.1

Consider the following information:

Proportion of acceptable components in a manufacturing batch, $y = 0.95$
Manufacturing rate of the facility in units per year, $P = 1250$
Demand rate in units per year, $D = 750$
Annual production-inventory carrying cost per unit, $I = ₹50$
Facility setup cost, $C = ₹75$

Now:
From the Eq. (3.4), optimal maximum production-inventory level can be obtained as:

$$V* = \sqrt{\frac{2CD(1-D/yP)}{I}}$$

$$= \sqrt{\frac{2 \times 75 \times 750 \times \{1 - 750/(0.95 \times 1250)\}}{50}}$$

$$= 28.79 \approx 29 \text{ units}$$

3.2.1 VARIATION OF MAXIMUM PRODUCTION-INVENTORY LEVEL

As mentioned before, the V value can be related to the space requirement and such requirement varies depending on the industry/product. However, for a generalized approach, V value serves the purpose. Table 3.1 shows the variation of V with respect to y value.

Reduction in the value of y indicates higher number of defective pieces in a manufacturing batch. This might be due to equipment malfunction, employment of less-skilled workers, wear and tear of tools, as well as alignment issues among other reasons faced by the company from time to time.

TABLE 3.1

Variation of V with respect to y

S. No.	Y	V
1	0.95	28.79
2	0.90	27.39
3	0.85	25.72
4	0.8	23.72
5	0.75	21.21

3.2.2 PERCENTAGE REDUCTION IN MAXIMUM INVENTORY LEVEL

For a generalized approach, let:

y_1 = Reduced value of y
V_1 = Reduced value of V

And:

$$V_1 = \sqrt{\frac{2CD(1 - D / y_1 P)}{I}}$$

Reduction in the maximum inventory level can be obtained as follows:

$$V - V_1 = \sqrt{\frac{2CD(1 - D / yP)}{I}} - \sqrt{\frac{2CD(1 - D / y_1 P)}{I}}$$

Or:

$$V - V_1 = \sqrt{\frac{2CD(1 - D / yP)}{I}} \left[1 - \sqrt{\frac{(1 - D / y_1 P)}{(1 - D / yP)}} \right]$$

Percentage reduction in V,

$$= \frac{(V - V_1)}{V} * 100$$

$$= \left[1 - \sqrt{\frac{(1 - D / y_1 P)}{(1 - D / yP)}} \right] * 100$$

Substituting the value of y_1 as:

$$y_1 = y \left(1 - \frac{W}{100} \right)$$

where W = % reduction in y.

TABLE 3.2

% Reduction in V with respect to % Reduction in y

S. No.	W	$\left[1-\sqrt{\dfrac{1-D/\{yP(1-W/100)\}}{(1-D/yP)}}\right]*100$
1	5	4.03
2	10	8.71
3	15	14.25
4	20	20.94
5	25	29.29

Now:

% reduction in V,

$$=\left[1-\sqrt{\frac{1-D/\{yP(1-W/100)\}}{(1-D/yP)}}\right]*100$$

Table 3.2 shows the % reduction in V with respect to the % reduction in y, for the given y, P, and D as follows:

Proportion of acceptable components in a manufacturing batch, y = 1.0
Manufacturing rate of the facility in units per year, P = 1250
Demand rate in units per year, D = 750

% reduction in V is more sensitive toward higher values of W.

Increase in the value of y indicates less number of defective pieces in a manu-facturing batch, i.e., an improvement in the quality practices. This might happen on account of better maintenance as well as setting of the facility, and employment of high-skilled worker among other reasons.

3.2.3 PERCENTAGE INCREASE IN MAXIMUM INVENTORY LEVEL

Increase in the value of y can result into an increase in the maximum production-inventory level in a production cycle. It needs to be studied in the context of quality criterion. This is because of an additional task related to arrangement of additional space requirement.

For a generalized approach, let:

y_1 = Increased value of y
V_1 = Increased value of V

And:

$$V_1=\sqrt{\frac{2CD(1-D/y_1P)}{I}}$$

Increase in the maximum inventory level can be obtained by subtracting V from V_1 as follows:

$$V_1 - V = \sqrt{\frac{2CD(1 - D / y_1 P)}{I}} - \sqrt{\frac{2CD(1 - D / yP)}{I}}$$

Or:

$$V_1 - V = \sqrt{\frac{2CD(1 - D / yP)}{I}} \left[\sqrt{\frac{(1 - D / y_1 P)}{(1 - D / yP)}} - 1 \right]$$

Percentage increase in V,

$$= \frac{(V_1 - V)}{V} * 100$$

$$= \left[\sqrt{\frac{(1 - D / y_1 P)}{(1 - D / yP)}} - 1 \right] * 100$$

Substituting the value of y_1 as:

$$y_1 = y \left(1 + \frac{W}{100} \right)$$

where W = % increase in y.

Now:

    % increase in V,

$$= \left[\sqrt{\frac{1 - D / \{yP(1 + W / 100)\}}{(1 - D / yP)}} - 1 \right] * 100$$

Table 3.3 shows the % increase in V with respect to the % increase in y, for the given y, P, and D as follows:

Proportion of acceptable components in a manufacturing batch, y = 0.8
Manufacturing rate of the facility in units per year, P = 1250
Demand rate in units per year, D = 750

TABLE 3.3

% Increase in V with respect to % Increase in y

S. No.	W	$\left[\sqrt{\dfrac{1 - D / \{yP(1 + W / 100)\}}{(1 - D / yP)}} - 1 \right] * 100$
1	5	6.90
2	10	12.81
3	15	17.95
4	20	22.47
5	25	26.49

From the Table 3.3, it can be observed that the % increase in V is less sensitive toward higher values of W.

3.3   EXAMPLES AND GENERALIZATION

In a business environment, different operational factors change with respect to time. Few examples along with the generalization are provided later in this text.

Example 3.2

With the following information:

Proportion of acceptable items, y = 0.8
Manufacturing rate, P = 1250
Demand rate, D = 750
Annual inventory carrying cost per unit, I = ₹50
Facility setup cost, C = ₹75

Value of maximum inventory level, V is obtained as equivalent to 23.72 units with the use of Eq. (3.4).

Now if the value of y increases to 0.9, the V value increases to 27.39.

With an increase in value of y, the maximum production-inventory level increases leading to certain additional space requirement. For an aim of bringing such requirement to an original level (i.e., for similar space), certain options need to be explored.

An option can be the reduction in rate of manufacture. Now:

$$V = \sqrt{\frac{2CD(1 - D/y_1 P_1)}{I}}$$

Or:

$$23.72 = \sqrt{\frac{2 \times 75 \times 750(1 - 750/0.9P_1)}{50}}$$

The reduced value of rate of manufacture, P_1 can be approximately evaluated as:

$$P_1 = 1111$$

In order to generalize, let:

$$y_1 = y\left(1 + \frac{W}{100}\right)$$

$$P_1 = P\left(1 - \frac{Z}{100}\right)$$

where:
  W = % increase in y
  Z = % reduction in P

For similar space requirement:

$$\sqrt{\frac{2CD(1 - D / yP)}{I}} = \sqrt{\frac{2CD(1 - D / y_1P_1)}{I}}$$

Or:

$$\frac{D}{yP} = \frac{D}{y_1P_1}$$

Or:

$$y_1P_1 = yP$$

Or:

$$P_1 = \frac{yP}{y_1}$$

Or:

$$P\left(1 - \frac{Z}{100}\right) = \frac{P}{(1 + W / 100)}$$

Or:

$$1 - \frac{Z}{100} = \frac{1}{(1 + W / 100)}$$

Or:

$$\frac{Z}{100} = 1 - \frac{1}{(1 + W / 100)}$$

Or:

$$\frac{Z}{100} = \frac{(W / 100)}{(1 + W / 100)}$$

Or:

$$Z = \frac{W}{(1 + W / 100)}$$

Table 3.4 shows the variation of Z with respect to W.
  Another option can be facility setup cost reduction.

TABLE 3.4
Variation of Z (Manufacturing Rate)
Corresponding to W(y)

S. No.	W	$Z = \dfrac{W}{(1 + W / 100)}$
1	5	4.76
2	10	9.09
3	15	13.04
4	20	16.67
5	25	20.00

Now:

$$y_1 = y\left(1 + \frac{W}{100}\right)$$

$$C_1 = C\left(1 - \frac{Z}{100}\right)$$

where:
   W = % increase in y
   Z = % reduction in C

For similar space requirement:

$$\sqrt{\frac{2CD(1 - D / yP)}{I}} = \sqrt{\frac{2C_1D(1 - D / y_1P)}{I}}$$

Or:

$$C(1 - D / yP) = C_1(1 - D / y_1P)$$

Or:

$$\frac{C_1}{C} = \frac{(1 - D / yP)}{(1 - D / y_1P)}$$

Or:

$$1 - \frac{Z}{100} = \frac{(1 - D / yP)}{(1 - D / y_1P)}$$

Or:

$$\frac{Z}{100} = 1 - \frac{(1 - D / yP)}{(1 - D / y_1P)}$$

TABLE 3.5

Variation of Z (Facility Setup Cost) Corresponding to W(y)

S. No.	W	$Z = \dfrac{DW}{yP(1+W/100)-D}$
1	5	12.50
2	10	21.43
3	15	28.13
4	20	33.33
5	25	37.50

Or:

$$\frac{Z}{100} = \frac{(D/yP)-(D/y_1P)}{(1-D/y_1P)}$$

Or:

$$\frac{Z}{100} = \frac{D(y_1/y)-D}{(y_1P)-D}$$

Or:

$$\frac{Z}{100} = \frac{D[(y_1/y)-1]}{(y_1P)-D}$$

Or:

$$\frac{Z}{100} = \frac{D(W/100)}{yP(1+W/100)-D}$$

Or:

$$Z = \frac{DW}{yP(1+W/100)-D}$$

Table 3.5 shows the variation of Z with respect to W, for the following values:

Proportion of acceptable components in a manufacturing batch, y = 0.8
Manufacturing rate of the facility in units per year, P = 1250
Demand rate in units per year, D = 750

Comparing with Table 3.4, Z values are much higher.

Example 3.3

In the previous example, the following options are explored:

i. Reduction in the rate of manufacture
ii. Reduction in the facility setup cost

In a practical situation, the manufacturing entrepreneur might select either the first or the second option depending on the degree of efforts needed and also the feasibility or implementation. However, both the options may also be implemented simultaneously.

From Table 3.5:

For $W = 10$; $Z = 21.43$

However, the entrepreneur feels that if the facility setup cost cannot be reduced by more than 15%, both the options can be implemented simultaneously.

In order to generalize:

$$y_1 = y\left(1 + \frac{W}{100}\right)$$

$$C_1 = C\left(1 - \frac{Z}{100}\right)$$

$$P_1 = P\left(1 - \frac{S}{100}\right)$$

where:
  W = % increase in y
  Z = % reduction in C
  S = % reduction in P

For similar maximum production-inventory level:

$$\sqrt{\frac{2CD(1 - D/yP)}{I}} = \sqrt{\frac{2C_1D(1 - D/y_1P_1)}{I}}$$

Or:

$$C(1 - D/yP) = C_1(1 - D/y_1P_1)$$

Or:

$$\left(1 - \frac{Z}{100}\right)\left[1 - \frac{D}{yP(1 + W/100)(1 - S/100)}\right] = (1 - D/yP)$$

Or:

$$1 - \frac{D}{yP(1 + W/100)(1 - S/100)} = \frac{(1 - D/yP)}{(1 - Z/100)}$$

Or:

$$\frac{D}{yP(1 + W/100)(1 - S/100)} = 1 - \frac{(1 - D/yP)}{(1 - Z/100)}$$

Or:

$$\frac{D}{yP(1+W/100)(1-S/100)} = \frac{1-(Z/100)-1+(D/yP)}{(1-Z/100)}$$

Or:

$$\frac{D}{yP(1+W/100)(1-S/100)} = \frac{(D/yP)-(Z/100)}{(1-Z/100)}$$

Or:

$$\frac{yP(1+W/100)(1-S/100)}{D} = \frac{(1-Z/100)}{(D/yP)-(Z/100)}$$

Or:

$$1-\frac{S}{100} = \frac{D(1-Z/100)}{yP(1+W/100)\{(D/yP)-(Z/100)\}}$$

Or:

$$\frac{S}{100} = 1-\frac{D(1-Z/100)}{yP(1+W/100)\{(D/yP)-(Z/100)\}}$$

Or:

$$S = 100\left[1-\frac{D(1-Z/100)}{yP(1+W/100)\{(D/yP)-(Z/100)\}}\right]$$

For the given values as follows:

Proportion of acceptable components in a manufacturing batch, y = 0.8
Manufacturing rate of the facility in units per year, P = 1250
Demand rate in units per year, D = 750

And for W = 10:

$$S = 100\left[1-\frac{750(1-Z/100)}{1100\{0.75-(Z/100)\}}\right]$$

Or:

$$S = 100\left[1-\frac{750-7.5Z}{825-11Z}\right]$$

Or:

$$S = 100\left[\frac{75-3.5Z}{825-11Z}\right]$$

TABLE 3.6
Combination of Z and S for W = 10

S. No.	Z	$S = 100\left[\dfrac{75 - 3.5Z}{825 - 11Z}\right]$
1	5	7.47
2	10	5.59
3	15	3.41
4	20	0.83

In case where Z cannot exceed 15, substituting $Z = 15$:

$$S = 3.41$$

For a particular set of operational factors, it is possible to generate various combinations of Z and S. In the present example, such combinations are provided in Table 3.6 for the value of W = 10. This helps with availability of wider choice in the context of manufacturing entrepreneurship.

Example 3.4

With the following information:

Proportion of acceptable items, $y = 0.8$
Manufacturing rate, $P = 1250$
Demand rate, $D = 750$
Annual inventory carrying cost per unit, $I = ₹50$
Facility setup cost, $C = ₹75$

Value of maximum inventory level, V is obtained as equivalent to 23.72 units with the use of Eq. (3.4).

Now if the facility setup cost is increased to ₹90 because of some reasons, the V value increases to 25.98.

For an aim of bringing space requirement to an original level (i.e., for similar space), suitable option should be explored.

A suitable option can be the reduction in P. Now:

$$V = \sqrt{\frac{2C_1 D(1 - D / yP_1)}{I}}$$

Or:

$$23.72 = \sqrt{\frac{2 \times 90 \times 750(1 - 750 / 0.8P_1)}{50}}$$

The reduced value of P, i.e., P_1 can be approximately evaluated as:

$$P_1 = 1184$$

In order to generalize, let:

$$C_1 = C\left(1 + \frac{W}{100}\right)$$

$$P_1 = P\left(1 - \frac{Z}{100}\right)$$

where:
  W = % increase in C
  Z = % reduction in P

For similar space requirement:

$$\sqrt{\frac{2CD(1 - D/yP)}{I}} = \sqrt{\frac{2C_1D(1 - D/yP_1)}{I}}$$

Or:

$$C(1 - D/yP) = C_1(1 - D/yP_1)$$

Or:

$$1 - \frac{D}{yP_1} = \frac{(1 - D/yP)}{(1 + W/100)}$$

Or:

$$\frac{D}{yP_1} = 1 - \frac{(1 - D/yP)}{(1 + W/100)}$$

Or:

$$\frac{D}{yP_1} = \frac{(W/100) + (D/yP)}{(1 + W/100)}$$

Or:

$$\frac{yP_1}{D} = \frac{(1 + W/100)}{(W/100) + (D/yP)}$$

Or:

$$\frac{yP}{D}\left(1 - \frac{Z}{100}\right) = \frac{(1 + W/100)}{(W/100) + (D/yP)}$$

Or:

$$1 - \frac{Z}{100} = \frac{(D/yP)(1 + W/100)}{(W/100) + (D/yP)}$$

TABLE 3.7

Variation of Z (Manufacturing Rate) Corresponding to W(C)

S. No.	W	$Z = \dfrac{W(1-D/yP)}{(W/100)+(D/yP)}$
1	5	1.56
2	10	2.94
3	15	4.17
4	20	5.26
5	25	6.25

Or:

$$\frac{Z}{100} = \frac{(W/100)+(D/yP)-(D/yP)-(D/yP)(W/100)}{(W/100)+(D/yP)}$$

Or:

$$\frac{Z}{100} = \frac{(W/100)(1-D/yP)}{(W/100)+(D/yP)}$$

Or:

$$Z = \frac{W(1-D/yP)}{(W/100)+(D/yP)}$$

Table 3.7 shows the variation of Z with respect to W for the given values.

Example 3.5

When annual inventory carrying cost per unit is reduced, the maximum production-inventory level increases. In order to bring back it to the original level, an option is the reduction in the facility setup cost.
Now:

$$I_1 = I\left(1-\frac{W}{100}\right)$$

$$C_1 = C\left(1-\frac{Z}{100}\right)$$

where:
  W = % reduction in I
  Z = % reduction in C

For similar maximum production-inventory level:

$$\sqrt{\frac{2CD(1-D/yP)}{I}} = \sqrt{\frac{2C_1D(1-D/yP)}{I_1}}$$

Or:

$$\frac{C}{I} = \frac{C_1}{I_1}$$

Or:

$$\frac{C_1}{C} = \frac{I_1}{I}$$

Or:

$$1 - \frac{Z}{100} = 1 - \frac{W}{100}$$

Or:

$$Z = W$$

Another option is the reduction in rate of manufacture. Now:

$$I_1 = I\left(1 - \frac{W}{100}\right)$$

$$P_1 = P\left(1 - \frac{Z}{100}\right)$$

where:
   W = % reduction in I
   Z = % reduction in P

And:

$$\sqrt{\frac{2CD(1 - D/yP)}{I}} = \sqrt{\frac{2CD(1 - D/yP_1)}{I_1}}$$

Or:

$$\frac{(1 - D/yP)}{I} = \frac{(1 - D/yP_1)}{I_1}$$

Or:

$$(I_1/I)(1 - D/yP) = (1 - D/yP_1)$$

Or:

$$\frac{D}{yP_1} = 1 - (I_1/I)(1 - D/yP)$$

Or:

$$\frac{yP_1}{D} = \frac{1}{1 - (I_1/I)(1 - D/yP)}$$

Or:

$$P_1 = \frac{(D/y)}{1-(I_1/I)(1-D/yP)}$$

Or:

$$1-\frac{Z}{100} = \frac{(D/yP)}{1-(1-W/100)(1-D/yP)}$$

Or:

$$\frac{Z}{100} = 1-\frac{(D/yP)}{1-(1-W/100)(1-D/yP)}$$

Or:

$$\frac{Z}{100} = \frac{1-(1-W/100)(1-D/yP)-(D/yP)}{1-(1-W/100)(1-D/yP)}$$

Or:

$$\frac{Z}{100} = \frac{1-1+(D/yP)+(W/100)-(W/100)(D/yP)-(D/yP)}{1-\{1-(D/yP)-(W/100)+(W/100)(D/yP)\}}$$

Or:

$$\frac{Z}{100} = \frac{(W/100)(1-D/yP)}{(D/yP)+(W/100)-(W/100)(D/yP)}$$

Or:

$$Z = \frac{W(1-D/yP)}{(D/yP)+(W/100)(1-D/yP)}$$

Table 3.8 shows the variation of Z with respect to W, for the given values as follows:

Proportion of acceptable components in a manufacturing batch, $y = 0.8$
Manufacturing rate of the facility in units per year, $P = 1250$
Demand rate in units per year, $D = 750$

TABLE 3.8

Variation of Z (Manufacturing Rate) Corresponding to W(I)

S. No.	W	$Z = \dfrac{W(1-D/yP)}{(D/yP)+(W/100)(1-D/yP)}$
1	5	1.64
2	10	3.23
3	15	4.76
4	20	6.25
5	25	7.69

While using the previous option, the values of Z are much higher because Z is equivalent to W for that option.

Example 3.6

In the previous example, the following options are explored:

  i. Reduction in the facility setup cost
  ii. Reduction in the rate of manufacture

In a practical situation, the manufacturing entrepreneur might select either the first or the second option depending on the degree of efforts needed as well as the feasibility or implementation. Quantitatively, the second option may appear favorable because values of Z are much lower. However, both the options may also be implemented simultaneously.
   Considering the first option:

   For W = 20; Z = 20

However, the entrepreneur might feel that if the facility setup cost cannot be reduced by more than 10%, both the options can be implemented simultaneously.
   In order to generalize:

$$I_1 = I\left(1 - \frac{W}{100}\right)$$

$$C_1 = C\left(1 - \frac{Z}{100}\right)$$

$$P_1 = P\left(1 - \frac{S}{100}\right)$$

where:
   W = % reduction in I
   Z = % reduction in C
   S = % reduction in P

For similar maximum production-inventory level:

$$\sqrt{\frac{2CD(1 - D/yP)}{I}} = \sqrt{\frac{2C_1D(1 - D/yP_1)}{I_1}}$$

Or:

$$\frac{C(1 - D/yP)}{I} = \frac{C_1(1 - D/yP_1)}{I_1}$$

Or:

$$1 - \frac{D}{yP_1} = \frac{CI_1(1 - D/yP)}{IC_1}$$

Or:

$$\frac{D}{yP_1} = \frac{IC_1 - CI_1(1 - D/yP)}{IC_1}$$

Or:

$$\frac{yP_1}{D} = \frac{IC_1}{IC_1 - CI_1(1 - D/yP)}$$

Or:

$$P_1 = \frac{(D/y)IC_1}{IC_1 - CI_1(1 - D/yP)}$$

Or:

$$1 - \frac{S}{100} = \frac{(D/yP)IC_1}{IC_1 - CI_1(1 - D/yP)}$$

Or:

$$\frac{S}{100} = \frac{IC_1 - CI_1(1 - D/yP) - (D/yP)IC_1}{IC_1 - CI_1(1 - D/yP)}$$

Or:

$$\frac{S}{100} = \frac{(C_1/C) - (I_1/I)(1 - D/yP) - (D/yP)(C_1/C)}{(C_1/C) - (I_1/I)(1 - D/yP)}$$

Or:

$$\frac{S}{100} = \frac{(C_1/C)(1 - D/yP) - (I_1/I)(1 - D/yP)}{(C_1/C) - (I_1/I)(1 - D/yP)}$$

Or:

$$\frac{S}{100} = \frac{(1 - D/yP)\{(C_1/C) - (I_1/I)\}}{(C_1/C) - (I_1/I)(1 - D/yP)}$$

Or:

$$\frac{S}{100} = \frac{(1 - D/yP)\{(1 - Z/100) - (1 - W/100)\}}{(1 - Z/100) - (1 - W/100)(1 - D/yP)}$$

Or:

$$\frac{S}{100} = \frac{(1 - D/yP)\{(W/100) - (Z/100)\}}{(1 - Z/100) - \{1 - (D/yP) - (W/100)(1 - D/yP)\}}$$

TABLE 3.9
Combination of Z and S for W = 20

S. No.	Z	$S = \dfrac{5-0.25Z}{0.8-(Z/100)}$
1	4	5.26
2	8	4.17
3	12	2.94
4	16	1.56

Or:

$$S = \frac{(1-D/yP)(W-Z)}{(D/yP)-(Z/100)+(W/100)(1-D/yP)}$$

For the given values as follows:

Proportion of acceptable components in a manufacturing batch, $y = 0.8$
Manufacturing rate of the facility in units per year, $P = 1250$
Demand rate in units per year, $D = 750$

And for W = 20:

$$S = \frac{5-0.25Z}{0.8-(Z/100)}$$

If Z cannot exceed 10, substituting $Z = 10$:

$$S = 3.57$$

For a particular set of operational factors, it is possible to generate various combinations of Z and S. In the present example, such combinations are provided in Table 3.9 for the value of W = 20. This helps in availability of wider choice in the context of manufacturing entrepreneurship.

Example 3.7

With the following information:

Proportion of acceptable items, $y = 0.8$
Manufacturing rate, $P = 1250$
Demand rate, $D = 750$
Annual inventory carrying cost per unit, $I = ₹50$
Facility setup cost, $C = ₹75$

Value of maximum inventory level, V is obtained as equivalent to 23.72 units with the use of Eq. (3.4).
Now if the value of P increases to 1500, the V value increases to 29.05.

With an increase in value of P, the maximum production-inventory level increases leading to certain additional space requirement. For an aim of bringing such requirement to an original level (i.e., for similar space), a suitable option can be a reduction in the facility setup cost. Now:

$$V = \sqrt{\frac{2C_1D(1-D/yP_1)}{I}}$$

Or:

$$23.72 = \sqrt{\frac{2 \times C_1 \times 750\{1 - 750/(0.8 \times 1500)\}}{50}}$$

The reduced value of facility setup cost, C_1 can be approximately evaluated as:

$$C_1 = ₹50$$

In order to generalize, let:

$$P_1 = P\left(1 + \frac{W}{100}\right)$$

$$C_1 = C\left(1 - \frac{Z}{100}\right)$$

where:
W = % increase in P
Z = % reduction in C

For similar space requirement:

$$\sqrt{\frac{2CD(1-D/yP)}{I}} = \sqrt{\frac{2C_1D(1-D/yP_1)}{I}}$$

Or:

$$C(1 - D/yP) = C_1(1 - D/yP_1)$$

Or:

$$\frac{C_1}{C} = \frac{(1-D/yP)}{(1-D/yP_1)}$$

Or:

$$1 - \frac{Z}{100} = \frac{(1-D/yP)}{(1-D/yP_1)}$$

Or:

$$\frac{Z}{100} = 1 - \frac{(1-D/yP)}{(1-D/yP_1)}$$

TABLE 3.10

Variation of Z (Facility Setup Cost) Corresponding to W(P)

S. No.	W	$Z = \dfrac{W}{(yP/D)(1+W/100)-1}$
1	5	12.50
2	10	21.43
3	15	28.12
4	20	33.33
5	25	37.5

Or:

$$\frac{Z}{100} = \frac{1-(D/yP_1)-1+(D/yP)}{(1-D/yP_1)}$$

Or:

$$\frac{Z}{100} = \frac{(D/yP)-(D/yP_1)}{(1-D/yP_1)}$$

Or:

$$\frac{Z}{100} = \frac{(P_1/P)-1}{(yP_1/D)-1}$$

Or:

$$\frac{Z}{100} = \frac{(1+W/100)-1}{(yP/D)(1+W/100)-1}$$

Or:

$$Z = \frac{W}{(yP/D)(1+W/100)-1}$$

Table 3.10 shows the variation of Z with respect to W for the given values.

An undesirable situation, when the proportion of acceptable components reduces in a lot, might be faced by an entrepreneur because of reasons beyond the control. Although the maximum production-inventory level may be less, however the space might be underutilized. In order to deal with such issues, few appropriate remedial measures can also be explored.

Example 3.8

With the following details:

Proportion of acceptable items, y = 1.0
Manufacturing rate, P = 1250

Demand rate, D = 750
Annual inventory carrying cost per unit, I = ₹50
Facility setup cost, C = ₹75

Value of maximum inventory level, V is obtained as equivalent to 30 units while using the formula:

$$V = \sqrt{\frac{2CD(1 - D/yP)}{I}}$$

When value of y reduces to y_1, a general approach is provided as follows:

$$y_1 = y\left(1 - \frac{W}{100}\right)$$

where W = % reduction in y.

One remedial measure can be the increase in manufacturing rate from P to P_1, i.e.,

$$P_1 = P\left(1 + \frac{Z}{100}\right)$$

where Z = % increase in P.

For similar maximum production-inventory level concerning the acceptable components to fulfill the demand:

$$\sqrt{\frac{2CD(1 - D/yP)}{I}} = \sqrt{\frac{2CD(1 - D/y_1P_1)}{I}}$$

Or:

$$yP = y_1P_1$$

Or:

$$\frac{P_1}{P} = \frac{y}{y_1}$$

Or:

$$1 + \frac{Z}{100} = \frac{1}{(1 - W/100)}$$

Or:

$$\frac{Z}{100} = \frac{1}{(1 - W/100)} - 1$$

Or:

$$\frac{Z}{100} = \frac{1 - 1 + (W/100)}{(1 - W/100)}$$

TABLE 3.11

Variation of Z (Manufacturing Rate)

Corresponding to W

S. No.	W	$Z = \dfrac{W}{(1-W/100)}$
1	5	5.26
2	10	11.11
3	15	17.65
4	20	25
5	25	33.33

Or:

$$\frac{Z}{100} = \frac{(W/100)}{(1-W/100)}$$

Or:

$$Z = \frac{W}{(1-W/100)}$$

Table 3.11 shows the variation of Z with respect to W.

Another remedial measure can be a reduction in the annual inventory carrying cost per unit.

Now:

$$\sqrt{\frac{2CD(1-D/yP)}{I}} = \sqrt{\frac{2CD(1-D/y_1P)}{I_1}}$$

Or:

$$\frac{1-D/yP}{I} = \frac{1-D/y_1P}{I_1}$$

Or:

$$\frac{I_1}{I} = \frac{1-D/y_1P}{1-D/yP}$$

Substituting,

$$I_1 = I\left(1 - \frac{Z}{100}\right)$$

$$1 - \frac{Z}{100} = \frac{1-D/y_1P}{1-D/yP}$$

Or:

$$\frac{Z}{100} = \frac{1-(D/yP)-1+(D/y_1P)}{1-(D/yP)}$$

Or:

$$\frac{Z}{100} = \frac{(D/y_1P)-(D/yP)}{1-(D/yP)}$$

Or:

$$\frac{Z}{100} = \frac{1-(y_1/y)}{(y_1P/D)-(y_1/y)}$$

Substituting,

$$y_1 = y\left(1-\frac{W}{100}\right)$$

$$\frac{Z}{100} = \frac{1-(1-W/100)}{(yP/D)(1-W/100)-(1-W/100)}$$

Or:

$$\frac{Z}{100} = \frac{(W/100)}{(1-W/100)\{(yP/D)-1\}}$$

Or:

$$Z = \frac{W}{(1-W/100)\{(yP/D)-1\}}$$

Table 3.12 shows the variation of Z with respect to W.
    Comparing with Table 3.11, the values of Z are relatively higher.

TABLE 3.12

Variation of Z (Inventory Holding Cost) Corresponding to W

S. No.	W	$Z = \dfrac{W}{(1-W/100)\{(yP/D)-1\}}$
1	5	7.89
2	10	16.65
3	15	26.48
4	20	30
5	25	50

Example 3.9

In the previous example, (i) manufacturing rate and (ii) inventory holding cost are varied as the remedial measures. For simultaneous variation:

$$y_1 = y\left(1 - \frac{W}{100}\right)$$

where W = % reduction in y.

$$P_1 = P\left(1 + \frac{Z}{100}\right)$$

where Z = % increase in P.

$$I_1 = I\left(1 - \frac{S}{100}\right)$$

where S = % reduction in I.
Now:

$$\sqrt{\frac{2CD(1 - D/yP)}{I}} = \sqrt{\frac{2CD(1 - D/y_1P_1)}{I_1}}$$

Or:

$$\frac{1 - D/yP}{I} = \frac{1 - D/y_1P_1}{I_1}$$

Or:

$$\frac{I_1}{I} = \frac{1 - D/y_1P_1}{1 - D/yP}$$

Or:

$$1 - \frac{S}{100} = \frac{1 - D/y_1P_1}{1 - D/yP}$$

Or:

$$\frac{S}{100} = \frac{1 - (D/yP) - 1 + (D/y_1P_1)}{1 - D/yP}$$

Or:

$$\frac{S}{100} = \frac{(D/y_1P_1) - (D/yP)}{1 - D/yP}$$

Or:

$$\frac{S}{100} = \frac{1 - (y_1P_1/yP)}{(y_1P_1/D) - (y_1P_1/yP)}$$

Or:

$$\frac{S}{100} = \frac{1-(1-W/100)(1+Z/100)}{(yP/D)(1-W/100)(1+Z/100)-(1-W/100)(1+Z/100)}$$

Or:

$$\frac{S}{100} = \frac{1-\left\{1+(Z/100)-(W/100)-(WZ/100^2)\right\}}{(1-W/100)(1+Z/100)\{(yP/D)-1\}}$$

Or:

$$\frac{S}{100} = \frac{(W/100)-(Z/100)+(WZ/100^2)}{(1-W/100)(1+Z/100)\{(yP/D)-1\}}$$

Or:

$$S = \frac{(W-Z)+(WZ/100)}{(1-W/100)(1+Z/100)\{(yP/D)-1\}}$$

For a particular set of operational factors, it is possible to generate various combinations of Z and S. In the present example, such combinations are provided in Table 3.13 for the value of W = 10. This helps in availability of wider choice in the context of manufacturing entrepreneurship. Depending on the ease in implementation, appropriate combination of Z and S can be chosen.

The formula used is for evaluation of S in a generalized way. Similarly, it is also possible to derive for evaluation of Z following a general approach.

Now:

$$\sqrt{\frac{2CD(1-D/yP)}{I}} = \sqrt{\frac{2CD(1-D/y_1P_1)}{I_1}}$$

Or:

$$\frac{1-D/yP}{I} = \frac{1-D/y_1P_1}{I_1}$$

Or:

$$1-(D/y_1P_1) = \frac{I_1(1-D/yP)}{I}$$

TABLE 3.13

Combination of Z and S for W = 10

S. No.	Z	$S = \dfrac{(W-Z)+(WZ/100)}{(1-W/100)(1+Z/100)\{(yP/D)-1\}}$
1	2	13.40
2	4	10.26
3	6	7.23
4	8	4.32

Or:

$$\frac{D}{y_1 P_1} = \frac{I - I_1(1 - D / yP)}{I}$$

Or:

$$\frac{y_1 P_1}{D} = \frac{I}{I - I_1(1 - D / yP)}$$

Or:

$$P_1 = \frac{(ID / y_1)}{I - I_1(1 - D / yP)}$$

Or:

$$1 + \frac{Z}{100} = \frac{(D / y_1 P)}{1 - (I_1 / I)(1 - D / yP)}$$

Or:

$$\frac{Z}{100} = \frac{(D / y_1 P) - 1 + (I_1 / I)(1 - D / yP)}{1 - (I_1 / I)(1 - D / yP)}$$

Or:

$$\frac{Z}{100} = \frac{(D / y_1 P) - 1 + (1 - S / 100)(1 - D / yP)}{1 - (1 - S / 100)(1 - D / yP)}$$

Or:

$$\frac{Z}{100} = \frac{(D / y_1 P) - 1 + 1 - (D / yP) - (S / 100) + (S / 100)(D / yP)}{1 - \{1 - (D / yP) - (S / 100) + (S / 100)(D / yP)\}}$$

Or:

$$\frac{Z}{100} = \frac{(D / y_1 P) - (D / yP) - (S / 100)(1 - D / yP)}{(D / yP) + (S / 100)(1 - D / yP)}$$

Or:

$$\frac{Z}{100} = \frac{1 - (y_1 / y) - (S / 100)\{(y_1 P / D) - (y_1 / y)\}}{(y_1 / y) + (S / 100)\{(y_1 P / D) - (y_1 / y)\}}$$

Or:

$$\frac{Z}{100} = \frac{1 - (1 - W / 100) - (S / 100)\{(yP / D)(1 - W / 100) - (1 - W / 100)\}}{(1 - W / 100) + (S / 100)\{(yP / D)(1 - W / 100) - (1 - W / 100)\}}$$

Or:

$$\frac{Z}{100} = \frac{1 - (1 - W / 100) - (1 - W / 100)(S / 100)\{(yP / D) - 1\}}{(1 - W / 100) + (1 - W / 100)(S / 100)\{(yP / D) - 1\}}$$

TABLE 3.14

Combination of S and Z for W = 10

S. No.	S	$Z = \dfrac{W - S(1 - W / 100)\{(yP / D) - 1\}}{(1 - W / 100)[1 + (S / 100)\{(yP / D) - 1\}]}$
1	2	9.65
2	4	7.55
3	6	6.84
4	8	5.48

Or:

$$\frac{Z}{100} = \frac{(W / 100) - (S / 100)(1 - W / 100)\{(yP / D) - 1\}}{(1 - W / 100)[1 + (S / 100)\{(yP / D) - 1\}]}$$

Or:

$$Z = \frac{W - S(1 - W / 100)\{(yP / D) - 1\}}{(1 - W / 100)[1 + (S / 100)\{(yP / D) - 1\}]}$$

In the present example, combinations of S and Z are provided in Table 3.14 for the value of W = 10.

Although the end customer demand is considered to depend on external factors. However, when it varies, its consideration at various interfaces within the organization is of interest for suitable generalization.

Example 3.10

If demand rate reduces, there is a possibility for requirement of additional space. A suitable option can be the reduction in rate of manufacture.

Now:

$$D_1 = D\left(1 - \frac{W}{100}\right)$$

where W = % reduction in D.

$$P_1 = P\left(1 - \frac{Z}{100}\right)$$

where Z = % reduction in P.

And:

$$\sqrt{\frac{2CD(1 - D / yP)}{I}} = \sqrt{\frac{2CD_1(1 - D_1 / yP_1)}{I}}$$

Or:

$$D(1 - D / yP) = D_1(1 - D_1 / yP_1)$$

Or:

$$(1 - D_1 / yP_1) = \frac{D(1 - D / yP)}{D_1}$$

Or:

$$\frac{D_1}{yP_1} = \frac{D_1 - D(1 - D / yP)}{D_1}$$

Or:

$$\frac{yP_1}{D_1} = \frac{D_1}{D_1 - D(1 - D / yP)}$$

Or:

$$P_1 = \frac{D_1^2}{y\{D_1 - D(1 - D / yP)\}}$$

Or:

$$1 - \frac{Z}{100} = \frac{D_1^2}{yP\{D_1 - D(1 - D / yP)\}}$$

Or:

$$1 - \frac{Z}{100} = \frac{D^2(1 - W / 100)^2}{yP\{D(1 - W / 100) - D(1 - D / yP)\}}$$

Or:

$$1 - \frac{Z}{100} = \frac{D(1 - W / 100)^2}{yP\{(1 - W / 100) - (1 - D / yP)\}}$$

Or:

$$1 - \frac{Z}{100} = \frac{D(1 - W / 100)^2}{yP\{(D / yP) - (W / 100)\}}$$

Or:

$$\frac{Z}{100} = 1 - \frac{D(1 - W / 100)^2}{D - (WyP / 100)}$$

Or:

$$Z = 100\left[1 - \frac{D(1 - W / 100)^2}{D - (WyP / 100)}\right]$$

Table 3.15 shows the variation of Z with respect to W, for the following values:

Proportion of acceptable components in a manufacturing batch, y = 0.8
Manufacturing rate of the facility in units per year, P = 1250
Demand rate in units per year, D = 750

TABLE 3.15

Variation of Z (Rate of Manufacture) Corresponding to W(D)

S. No.	W	$Z = 100\left[1 - \dfrac{D(1-W/100)^2}{D-(WyP/100)}\right]$
1	5	3.30
2	10	6.54
3	15	9.69
4	20	12.73
5	25	15.62

Another option can be the facility setup cost reduction.
Now:

$$D_1 = D\left(1 - \frac{W}{100}\right)$$

where W = % reduction in D.

$$C_1 = C\left(1 - \frac{Z}{100}\right)$$

where Z = % reduction in C.
And:

$$\sqrt{\frac{2CD(1-D/yP)}{I}} = \sqrt{\frac{2C_1 D_1(1-D_1/yP)}{I}}$$

Or:

$$CD(1-D/yP) = C_1 D_1(1-D_1/yP)$$

Or:

$$(1-D/yP) = (1-Z/100)(1-W/100)\{1-(D/yP)(1-W/100)\}$$

Or:

$$1 - \frac{Z}{100} = \frac{(1-D/yP)}{(1-W/100)\{1-(D/yP)(1-W/100)\}}$$

Or:

$$\frac{Z}{100} = 1 - \frac{(1-D/yP)}{(1-W/100)\{1-(D/yP)(1-W/100)\}}$$

TABLE 3.16

Variation of Z (Facility Setup Cost) Corresponding to W(D)

S. No.	W	$Z = 100\left[1 - \dfrac{(1-D/yP)}{(1-W/100)\{1-(D/yP)(1-W/100)\}}\right]$
1	5	8.47
2	10	14.53
3	15	18.86
4	20	21.87
5	25	23.81

Or:

$$Z = 100\left[1 - \frac{(1-D/yP)}{(1-W/100)\{1-(D/yP)(1-W/100)\}}\right]$$

Table 3.16 shows the variation of Z with respect to W, for the given values.

Comparing with Table 3.15, the values of Z are much higher.
For simultaneous variation of both the options:

$$D_1 = D\left(1 - \frac{W}{100}\right)$$

$$P_1 = P\left(1 - \frac{Z}{100}\right)$$

$$C_1 = C\left(1 - \frac{S}{100}\right)$$

Now:

$$\sqrt{\frac{2CD(1-D/yP)}{I}} = \sqrt{\frac{2C_1D_1(1-D_1/yP_1)}{I}}$$

Or:

$$CD(1-D/yP) = C_1D_1(1-D_1/yP_1)$$

Or:

$$(1-D/yP) = (1-S/100)(1-W/100)\{1-(D/yP)(1-W/100)/(1-Z/100)\}$$

Or:

$$1 - \frac{S}{100} = \frac{(1-D/yP)}{(1-W/100)\{1-(D/yP)(1-W/100)/(1-Z/100)\}}$$

Or:

$$\frac{S}{100} = 1 - \frac{(1-D/yP)}{(1-W/100)\{1-(D/yP)(1-W/100)/(1-Z/100)\}}$$

TABLE 3.17

Combination of Z and S for W = 20

S. No.	Z	$S = 100\left[1 - \dfrac{(1 - D/yP)}{(1 - W/100)\{1 - (D/yP)(1 - W/100)/(1 - Z/100)\}}\right]$
1	2	19.41
2	4	16.67
3	6	13.60
4	8	10.16

Or:

$$S = 100\left[1 - \frac{(1 - D/yP)}{(1 - W/100)\{1 - (D/yP)(1 - W/100)/(1 - Z/100)\}}\right]$$

In the present example, few combinations of Z and S are provided in Table 3.17 for the value of W = 20, considering following details:

Proportion of acceptable components in a manufacturing batch, y = 0.8
Manufacturing rate of the facility in units per year, P = 1250
Demand rate in units per year, D = 750
Annual inventory carrying cost per unit, I = ₹50
Facility setup cost, C = ₹75

Example 3.11

In case of the demand rate increase, the following options may be exercised for better space utilization:

    i. Increase in y
   ii. Reduction in I
  iii. Increase in P

With the first option:

$$D_1 = D\left(1 + \frac{W}{100}\right)$$

$$y_1 = y\left(1 + \frac{Z}{100}\right)$$

$$\sqrt{\frac{2CD(1 - D/yP)}{I}} = \sqrt{\frac{2CD_1(1 - D_1/y_1P)}{I}}$$

Or:

$$D(1 - D/yP) = D_1(1 - D_1/y_1P)$$

Or:

$$1 - \frac{D_1}{y_1 P} = \frac{D(1 - D/yP)}{D_1}$$

Or:

$$1 - \frac{D_1}{y_1 P} = \frac{(1 - D/yP)}{(1 + W/100)}$$

Or:

$$\frac{D_1}{y_1 P} = \frac{1 + (W/100) - 1 + (D/yP)}{(1 + W/100)}$$

Or:

$$\frac{y_1 P}{D_1} = \frac{(1 + W/100)}{(W/100) + (D/yP)}$$

Or:

$$y_1 = \frac{(D_1/P)(1 + W/100)}{(W/100) + (D/yP)}$$

Or:

$$1 + \frac{Z}{100} = \frac{(D_1/yP)(1 + W/100)}{(W/100) + (D/yP)}$$

Or:

$$\frac{Z}{100} = \frac{(D/yP)(1 + W/100)^2}{(W/100) + (D/yP)} - 1$$

Or:

$$Z = 100 \left[\frac{(D/yP)(1 + W/100)^2}{(W/100) + (D/yP)} - 1 \right]$$

Table 3.18 shows the variation of Z with respect to W, for the following values:

Proportion of acceptable components in a manufacturing batch, y = 0.8
Manufacturing rate of the facility in units per year, P = 1250
Demand rate in units per year, D = 750

TABLE 3.18
Variation of Z(Y) Corresponding to W (Demand Rate)

S. No.	W	$Z = 100 \left[\dfrac{(D/yP)(1 + W/100)^2}{(W/100) + (D/yP)} - 1 \right]$
1	5	3.36
2	10	6.76
3	15	10.21
4	20	13.68

With the second option:

$$D_1 = D\left(1 + \frac{W}{100}\right)$$

$$I_1 = I\left(1 - \frac{Z}{100}\right)$$

$$\sqrt{\frac{2CD(1 - D/yP)}{I}} = \sqrt{\frac{2CD_1(1 - D_1/yP)}{I_1}}$$

Or:

$$\frac{D(1 - D/yP)}{I} = \frac{D_1(1 - D_1/yP)}{I_1}$$

Or:

$$1 - \frac{Z}{100} = \frac{D_1(1 - D_1/yP)}{D(1 - D/yP)}$$

Or:

$$\frac{Z}{100} = 1 - \frac{(1 + W/100)\{1 - (D/yP)(1 + W/100)\}}{(1 - D/yP)}$$

Or:

$$Z = 100\left[1 - \frac{(1 + W/100)\{1 - (D/yP)(1 + W/100)\}}{(1 - D/yP)}\right]$$

Variation of Z corresponding to W is represented by Table 3.19 for the given input factors.

With the third option:

$$D_1 = D\left(1 + \frac{W}{100}\right)$$

$$P_1 = P\left(1 + \frac{Z}{100}\right)$$

$$\sqrt{\frac{2CD(1 - D/yP)}{I}} = \sqrt{\frac{2CD_1(1 - D_1/yP_1)}{I}}$$

TABLE 3.19

Variation of Z(I) Corresponding to W (Demand Rate)

S. No.	W	$Z = 100\left[1 - \dfrac{(1 + W/100)\{1 - (D/yP)(1 + W/100)\}}{(1 - D/yP)}\right]$
1	5	10.75
2	10	23.00
3	15	36.75
4	20	52.00

Or:

$$D(1-D/yP) = D_1(1-D_1/yP_1)$$

Following similar procedure, the Z value can be derived as:

$$Z = 100\left[\frac{(D/yP)(1+W/100)^2}{(W/100)+(D/yP)}-1\right]$$

With the combination of options, consider y and I. Now:

$$D_1 = D\left(1+\frac{W}{100}\right)$$

$$y_1 = y\left(1+\frac{Z}{100}\right)$$

$$I_1 = I\left(1-\frac{S}{100}\right)$$

$$\sqrt{\frac{2CD(1-D/yP)}{I}} = \sqrt{\frac{2CD_1(1-D_1/y_1P)}{I_1}}$$

Or:

$$\frac{D(1-D/yP)}{I} = \frac{D_1(1-D_1/y_1P)}{I_1}$$

Or:

$$\frac{I_1}{I} = \frac{D_1(1-D_1/y_1P)}{D(1-D/yP)}$$

Or:

$$1-\frac{S}{100} = \frac{(1+W/100)(1-D_1/y_1P)}{(1-D/yP)}$$

Or:

$$\frac{S}{100} = 1 - \frac{(1+W/100)\{1-(D/yP)(1+W/100)/(1+Z/100)\}}{(1-D/yP)}$$

Or:

$$S = 100\left[1 - \frac{(1+W/100)\{1-(D/yP)(1+W/100)/(1+Z/100)\}}{(1-D/yP)}\right]$$

In the present example, certain combinations of Z and S are represented in Table 3.20 for W = 10.

TABLE 3.20
Combination of Z and S for W = 10

S. No.	Z	$S = 100\left[1 - \dfrac{(1+W/100)\{1-(D/yP)(1+W/100)/(1+Z/100)\}}{(1-D/yP)}\right]$
1	2	15.88
2	3	12.43
3	4	9.04
4	5	5.71

Now, consider I and P for the combination of options.

$$D_1 = D\left(1+\frac{W}{100}\right)$$

$$I_1 = I\left(1-\frac{Z}{100}\right)$$

$$P_1 = P\left(1+\frac{S}{100}\right)$$

$$\sqrt{\frac{2CD(1-D/yP)}{I}} = \sqrt{\frac{2CD_1(1-D_1/yP_1)}{I_1}}$$

Or:

$$\frac{D(1-D/yP)}{I} = \frac{D_1(1-D_1/yP_1)}{I_1}$$

Or:

$$1 - \frac{D_1}{yP_1} = \frac{(1-Z/100)(1-D/yP)}{(1+W/100)}$$

Or:

$$\frac{D_1}{yP_1} = \frac{(1+W/100)-(1-Z/100)(1-D/yP)}{(1+W/100)}$$

Or:

$$\frac{yP_1}{D_1} = \frac{(1+W/100)}{(1+W/100)-(1-Z/100)(1-D/yP)}$$

Or:

$$P_1 = \frac{(D_1/y)(1+W/100)}{(1+W/100)-(1-Z/100)(1-D/yP)}$$

TABLE 3.21

Combination of Z and S for W = 20

S. No.	Z	$S = 100\left[\dfrac{(D/yP)(1+W/100)^2}{(1+W/100)-(1-Z/100)(1-D/yP)} - 1\right]$
1	4	12.50
2	8	11.34
3	12	10.20
4	16	9.09

Or:

$$1 + \frac{S}{100} = \frac{(D_1/yP)(1+W/100)}{(1+W/100)-(1-Z/100)(1-D/yP)}$$

Or:

$$\frac{S}{100} = \frac{(D/yP)(1+W/100)^2}{(1+W/100)-(1-Z/100)(1-D/yP)} - 1$$

Or:

$$S = 100\left[\frac{(D/yP)(1+W/100)^2}{(1+W/100)-(1-Z/100)(1-D/yP)} - 1\right]$$

In the present example, certain combinations of Z and S are represented in Table 3.21 for W = 20.

After providing a basic understanding of the quality aspects and also mathematical treatment, many examples along with the generalization have been rigorously analyzed. This is in reference to the fluctuation in operational factors and appropriate response to such aspects with specific inclusion of quality criterion.

4 Backorders

A successor facility needs the output of a predecessor facility. This need of the successor facility may be visualized as its demand. Generally speaking also, the demand may be:

i. either completely fulfilled; or
ii. not fulfilled completely.

Demand may not be fulfilled completely now, because:

a. There might be maintenance problems with the predecessor facility.
b. There might be human resources issue, in terms of absence.
c. Non-availability of input item is observed for the predecessor facility.
d. Quality problems are observed concerning the predecessor facility.

If the demand is not fulfilled completely now, the shortfall may be backordered in the manufacturing concern. This leads to backorders in the system and should be understood in the context of manufacturing entrepreneurship.

Level of demand for the successor facility is represented by Figure 4.1. This may be visualized as almost full container/box placed before the successor facility.

However, the fulfilled demand in terms of manufactured quantity from the predecessor facility is represented by Figure 4.2. This is not equivalent to the desired level of demand of the successor facility, but less than that. This shortfall may be due to the reasons mentioned before, i.e., in terms of availability of resources, quality, and repair issues among other problems being faced by the organizations from time to time, in their journey for completion of manufacturing orders. This shortfall in the demand of the successor facility indicates the backorders.

The discussed shortfall needs to be backordered. The shortages which are backordered can also be included for the analysis of maximum inventory level under such a scenario.

4.1 MAXIMUM INVENTORY LEVEL

In order to find out the maximum inventory level during the manufacturing cycle, refer to Figure 4.3 along with the backorders, where:

$0-t_1$ is the time during which production activity for backorders happens
t_2-t_3 is the time during which shortages are faced, which are completely backordered
P = Manufacturing rate in units per year
D = Demand rate in units per year

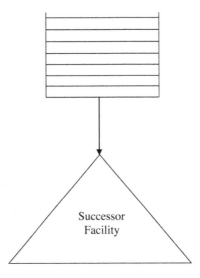

FIGURE 4.1   Representation for level of demand.

V = Maximum production-inventory level
J = Maximum backordering quantity

Now:
Manufacturing cycle time,

$$T = \frac{(V+J)}{(P-D)} + \frac{(V+J)}{D}$$

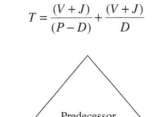

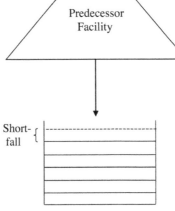

FIGURE 4.2   Shortfall in the desired demand indicating the backorders.

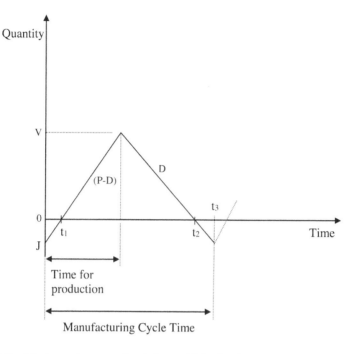

FIGURE 4.3   Maximum inventory level along with backorders.

Or:

$$T = (V + J)\left[\frac{D + P - D)}{D(P - D)}\right]$$

Or:

$$T = \frac{P(V + J)}{D(P - D)}$$

Or:

$$T = \frac{(V + J)}{D(1 - D / P)} \qquad (4.1)$$

And: Number of cycles in a year = 1/(Cycle time, T)
Or:
   Number of cycles in a year,

$$= \frac{D(1 - D / P)}{(V + J)}$$

Since facility setup cost is incurred for each cycle, an annual facility setup cost is evaluated as the multiplication of such setup cost and the number of cycles in a year. Therefore:

Annual facility setup cost,

$$= \frac{CD(1 - D/P)}{(V + J)} \qquad (4.2)$$

where C = Facility setup cost.

As the average inventory is:

$$\frac{V}{2}$$

And positive inventory exists in a cycle for the period:

$$\frac{V}{(P - D)} + \frac{V}{D}$$

Annual production-inventory carrying cost can be evaluated as:

$$\frac{VI}{2} \left[\frac{V}{(P - D)} + \frac{V}{D} \right] * \text{Number of cycles in a year}$$

where I = Annual production-inventory carrying cost per unit.

Now:

Annual production-inventory carrying cost,

$$= \frac{VI}{2} \left[\frac{V}{(P - D)} + \frac{V}{D} \right] \frac{D(1 - D/P)}{(V + J)} \qquad (4.3)$$

Similarly, period corresponding to backorders,

$$= \frac{J}{(P - D)} + \frac{J}{D}$$

And, average backordering quantity,

$$= \frac{J}{2}$$

Also with, the number of cycles in a year,

$$= \frac{D(1 - D/P)}{(V + J)}$$

The annual backordering cost,

$$= \frac{JK}{2} \left[\frac{J}{(P - D)} + \frac{J}{D} \right] \frac{D(1 - D/P)}{(V + J)} \qquad (4.4)$$

where K = Annual backordering cost per unit.

Adding the expressions (4.2), (4.3), and (4.4), the total related cost (E) is formulated as:

$$E = \frac{CD(1 - D/P)}{(V + J)} + \frac{VI}{2} \left[\frac{V}{(P - D)} + \frac{V}{D} \right] \frac{D(1 - D/P)}{(V + J)} + \frac{JK}{2} \left[\frac{J}{(P - D)} + \frac{J}{D} \right] \frac{D(1 - D/P)}{(V + J)}$$

Or:

$$E = \frac{CD(1-D/P)}{(V+J)} + \frac{V^2I}{2}\left[\frac{D+P-D}{D(P-D)}\right]\frac{D(1-D/P)}{(V+J)} + \frac{J^2K}{2}\left[\frac{D+P-D}{D(P-D)}\right]\frac{D(1-D/P)}{(V+J)}$$

Or:

$$E = \frac{CD(1-D/P)}{(V+J)} + \frac{V^2I}{2}\left[\frac{1}{D(1-D/P)}\right]\frac{D(1-D/P)}{(V+J)} + \frac{J^2K}{2}\left[\frac{1}{D(1-D/P)}\right]\frac{D(1-D/P)}{(V+J)}$$

Or:

$$E = \frac{CD(1-D/P)}{(V+J)} + \frac{V^2I}{2(V+J)} + \frac{J^2K}{2(V+J)} \quad (4.5)$$

From Eq. (4.1):

$$V + J = TD(1-D/P) \quad (4.6)$$

Substituting in Eq. (4.5):

$$E = \frac{C}{T} + \frac{V^2I}{2TD(1-D/P)} + \frac{J^2K}{2TD(1-D/P)}$$

Using the value of J from Eq. (4.6)

$$E = \frac{C}{T} + \frac{V^2I + K[TD(1-D/P)-V]^2}{2TD(1-D/P)}$$

Or:

$$E = \frac{C}{T} + \frac{V^2I + K[T^2D^2(1-D/P)^2 - 2TDV(1-D/P)+V^2]}{2TD(1-D/P)}$$

Or:

$$E = \frac{C}{T} + \frac{V^2(I+K) + KTD(1-D/P)[TD(1-D/P)-2V]}{2TD(1-D/P)}$$

Or:

$$E = \frac{C}{T} + \frac{V^2(I+K)}{2TD(1-D/P)} + \frac{KTD(1-D/P)}{2} - VK \quad (4.7)$$

With an objective of minimizing the total related cost, differentiating partially with respect to V and equating to zero,

$$\frac{V(I+K)}{TD(1-D/P)} = K$$

Or:

$$TD(1-D/P) = \frac{V(I+K)}{K}$$

Substituting in Eq. (4.7),

$$E = \frac{CKD(1 - D/P)}{V(I+K)} + \frac{V^2(I+K)K}{2V(I+K)} + \frac{KV(I+K)}{2K} - VK$$

Or:

$$E = \frac{CKD(1 - D/P)}{V(I+K)} + \frac{VK}{2} + \frac{V(I+K)}{2} - VK$$

Or:

$$E = \frac{CKD(1 - D/P)}{V(I+K)} + \frac{VI}{2} \tag{4.8}$$

Differentiating with respect to V and equating to zero,

$$\frac{I}{2} - \frac{CKD(1 - D/P)}{V^2(I+K)} = 0$$

Or:

$$V^2 = \frac{2CKD(1 - D/P)}{I(I+K)}$$

Or:

$$V* = \sqrt{\frac{2CKD(1 - D/P)}{I(I+K)}} \tag{4.9}$$

From the above expression:

$$\sqrt{\frac{2CKD(1 - D/P)}{(I+K)}} = V* \sqrt{I}$$

Substituting in Eq. (4.8),

$$E* = \frac{IV*}{2} + \frac{IV*}{2}$$

Or:

$$E* = IV* \tag{4.10}$$

Example 4.1

Consider the following information:

> Manufacturing rate of the predecessor facility in units per year, P = 1250
> Demand rate of the successor facility in units per year, D = 750
> Annual production-inventory carrying cost per unit, I = ₹50
> Facility setup cost, C = ₹75
> Annual backordering cost per unit, K = ₹100

Now:

From the Eq. (4.9), optimal maximum inventory level can be obtained as:

$$V^* = \sqrt{\frac{2CKD(1-D/P)}{I(I+K)}}$$

$$= \sqrt{\frac{2 \times 75 \times 100 \times 750 \times (1-750/1250)}{50(50+100)}}$$

$$= 24.49 \text{ units}$$

From the Eq. (4.10), total related cost is obtained as follows:

$$E^* = IV^*$$

$$= 50 \times 24.49$$

$$= ₹1224.74$$

4.1.1 INCREASE IN MAXIMUM INVENTORY LEVEL

With an increase in the backordering cost, the variation in maximum inventory level is illustrated with the following example.

Example 4.2

With the data of previous example, if the annual backordering cost per unit increases by 10%, then the revised V value can be obtained as follows:

$$V = \sqrt{\frac{2CKD(1-D/P)}{I(I+K)}}$$

$$= \sqrt{\frac{2 \times 75 \times 110 \times 750 \times (1-750/1250)}{50(50+110)}}$$

$$= 24.87 \text{ units}$$

That is approximately 1.55% increase in V value.

For a general approach:
Increase in the maximum inventory level,

$$V_1 - V = \sqrt{\frac{2CK_1D(1-D/P)}{I(I+K_1)}} - \sqrt{\frac{2CKD(1-D/P)}{I(I+K)}}$$

where K_1 = Increased annual backordering cost per unit.

And V_1 = Corresponding increased maximum inventory level.

Now:

$$V_1 - V = \sqrt{\frac{2CKD(1-D/P)}{I(I+K)}}\left[\sqrt{\frac{(I+K)K_1}{K(I+K_1)}} - 1\right]$$

TABLE 4.1

% Increase in V with respect to % Increase in K

S. No.	W	$\left[\sqrt{\dfrac{(I+K)(1+W/100)}{I+K(1+W/100)}} - 1\right] * 100$
1	10	1.55
2	20	2.90
3	30	4.08
4	40	5.13
5	50	6.07

Percentage increase in V,

$$= \frac{(V_1 - V)}{V} * 100$$

$$= \left[\sqrt{\frac{(I+K)K_1}{K(I+K_1)}} - 1\right] * 100$$

Substituting the value of K_1 as:

$$K_1 = K\left(1 + \frac{W}{100}\right)$$

where W = % increase in K.

Now:

% increase in V,

$$= \left[\sqrt{\frac{(I+K)(1+W/100)}{I+K(1+W/100)}} - 1\right] * 100$$

Using the given data, Table 4.1 shows the % increase in V with respect to the % increase in K.

4.1.2 REDUCTION IN MAXIMUM INVENTORY LEVEL

If the annual backordering cost per unit, K reduces, the reduction in the maximum inventory level,

$$V - V_1 = \sqrt{\frac{2CKD(1-D/P)}{I(I+K)}} - \sqrt{\frac{2CK_1D(1-D/P)}{I(I+K_1)}}$$

where K_1 = Reduced value of K.

And V_1 = Corresponding reduced maximum inventory level

Now:

$$V - V_1 = \sqrt{\frac{2CKD(1-D/P)}{I(I+K)}}\left[1 - \sqrt{\frac{(I+K)K_1}{K(I+K_1)}}\right]$$

TABLE 4.2
% Reduction in V with respect to % Reduction in K

S. No.	W	$\left[1-\sqrt{\dfrac{(I+K)(1-W/100)}{I+K(1-W/100)}}\right]*100$
1	10	1.80
2	20	3.92
3	30	6.46
4	40	9.55
5	50	13.40

% reduction in V,

$$= \frac{(V - V_1)}{V} * 100$$

$$= \left[1 - \sqrt{\frac{(I+K)K_1}{K(I+K_1)}}\right] * 100$$

Substituting the value of K_1 as:

$$K_1 = K\left(1 - \frac{W}{100}\right)$$

where $W = \%$ reduction in K.

Now:

% reduction in V,

$$= \left[1 - \sqrt{\frac{(I+K)(1-W/100)}{I+K(1-W/100)}}\right] * 100$$

Using the given relevant data such as:

Annual production-inventory carrying cost per unit, $I = ₹50$
Annual backordering cost per unit, $K = ₹100$

The above-mentioned formula can be implemented for evaluation of % reduction in V with respect to the % decrease in K value. Table 4.2 shows the % reduction in maximum production-inventory level with respect to the % reduction in the annual backordering cost per unit item. The values are more sensitive toward higher values of W. On comparison with Table 4.1, the % variation is higher.

Example 4.3

Consider the reduction in annual production-inventory carrying cost per unit (I), and associated increase in V.

Increase in the maximum inventory level,

$$V_1 - V = \sqrt{\frac{2CKD(1-D/P)}{I_1(I_1+K)}} - \sqrt{\frac{2CKD(1-D/P)}{I(I+K)}}$$

where I_1 = Increased value of I.

And V_1 = Corresponding increased maximum inventory level.

Now:

$$V_1 - V = \sqrt{\frac{2CKD(1-D/P)}{I(I+K)}}\left[\sqrt{\frac{I(I+K)}{I_1(I_1+K)}} - 1\right]$$

% increase in V,

$$= \frac{(V_1-V)}{V} * 100$$

$$= \left[\sqrt{\frac{I(I+K)}{I_1(I_1+K)}} - 1\right] * 100$$

Substituting the value of I_1 as:

$$I_1 = I\left(1 - \frac{W}{100}\right)$$

where W = % reduction in I.

Now:
  % increase in V,

$$= \left[\sqrt{\frac{(I+K)}{(1-W/100)\{I(1-W/100)+K\}}} - 1\right] * 100$$

With the use of relevant data such as:

Annual production-inventory carrying cost per unit, I = ₹50
Annual backordering cost per unit, K = ₹100

Table 4.3 shows the % increase in V with respect to the % reduction in I. W is varied from 10 to 50 with the increment of 10, and the corresponding values of % increase in V are obtained. The obtained values are more sensitive as W increases.
    Now, with the increase in annual production-inventory carrying cost per unit (I), the associated reduction in V is as follows:

$$V - V_1 = \sqrt{\frac{2CKD(1-D/P)}{I(I+K)}} - \sqrt{\frac{2CKD(1-D/P)}{I_1(I_1+K)}}$$

TABLE 4.3

% Increase in V with respect to % Reduction in I

S. No.	W	$\left[\sqrt{\dfrac{(I+K)}{(1-W/100)\{I(1-W/100)+K\}}}-1\right]*100$
1	10	7.21
2	20	15.73
3	30	25.99
4	40	38.68
5	50	54.92

Or:

$$V - V_1 = \sqrt{\frac{2CKD(1-D/P)}{I(I+K)}}\left[1-\sqrt{\frac{I(I+K)}{I_1(I_1+K)}}\right]$$

% reduction in V,

$$= \frac{(V-V_1)}{V}*100$$

$$= \left[1-\sqrt{\frac{I(I+K)}{I_1(I_1+K)}}\right]*100$$

Substituting the value of I_1 as:

$$I_1 = I\left(1+\frac{W}{100}\right)$$

where W = % increase in I.

Now:

% reduction in V,

$$= \left[1-\sqrt{\frac{(I+K)}{(1+W/100)\{I(1+W/100)+K\}}}\right]*100$$

With the use of given data, Table 4.4 shows the % reduction in V with respect to the % increase in I. The obtained values are less sensitive as W increases. In comparison with Table 4.3, the % variation in V is lower.

TABLE 4.4

% Reduction in V with respect to % Increase in I

S. No.	W	$\left[1-\sqrt{\dfrac{(I+K)}{(1+W/100)\{I(1+W/100)+K\}}}\right]*100$
1	10	6.20
2	20	11.61
3	30	16.37
4	40	20.61
5	50	24.41

Example 4.4

Consider the increase in manufacturing rate (P), and associated increase in V.
Increase in the maximum inventory level,

$$V_1 - V = \sqrt{\frac{2CKD(1 - D/P_1)}{I(I + K)}} - \sqrt{\frac{2CKD(1 - D/P)}{I(I + K)}}$$

where P_1 = Increased value of P.

And V_1 = Corresponding increased maximum inventory level.

Now:

$$V_1 - V = \sqrt{\frac{2CKD(1 - D/P)}{I(I + K)}} \left[\sqrt{\frac{(1 - D/P_1)}{(1 - D/P)}} - 1 \right]$$

Percentage increase in V,

$$= \frac{(V_1 - V)}{V} * 100$$

$$= \left[\sqrt{\frac{(1 - D/P_1)}{(1 - D/P)}} - 1 \right] * 100$$

Substituting the value of P_1 as:

$$P_1 = P\left(1 + \frac{W}{100}\right)$$

where W = % increase in P.

Now:
% increase in V,

$$= \left[\sqrt{\frac{\{1 - D/P(1 + W/100)\}}{(1 - D/P)}} - 1 \right] * 100$$

With the use of relevant data such as:

Manufacturing rate of the predecessor facility in units per year, P = 1250
Demand rate of the successor facility in units per year, D = 750

Table 4.5 shows the % increase in V with respect to the % increase in P.
Now, with the reduction in manufacturing rate (P), the associated reduction in V is as follows:

$$V - V_1 = \sqrt{\frac{2CKD(1 - D/P)}{I(I + K)}} - \sqrt{\frac{2CKD(1 - D/P_1)}{I(I + K)}}$$

TABLE 4.5
% Increase in V with respect to % Increase in P

S. No.	W	$\left[\sqrt{\dfrac{\{1-D/P(1+W/100)\}}{(1-D/P)}}-1\right]*100$
1	5	3.51
2	10	6.60
3	15	9.35
4	20	11.80
5	25	14.02

TABLE 4.6
% Reduction in V with respect to % Reduction in P

S. No.	W	$\left[1-\sqrt{\dfrac{\{1-D/P(1-W/100)\}}{(1-D/P)}}\right]*100$
1	5	4.03
2	10	8.71
3	15	14.25
4	20	20.94
5	25	29.29

Or:

$$V-V_1 = \sqrt{\frac{2CKD(1-D/P)}{I(I+K)}}\left[1-\sqrt{\frac{\{1-D/P(1-W/100)\}}{(1-D/P)}}\right]$$

substituting the value of P_1 as:

$$P_1 = P\left(1-\frac{W}{100}\right)$$

% reduction in V,

$$= \frac{(V-V_1)}{V}*100$$

$$= \left[1-\sqrt{\frac{\{1-D/P(1-W/100)\}}{(1-D/P)}}\right]*100$$

Table 4.6 shows the % reduction in V with respect to % reduction in P. In comparison with Table 4.5, the values are higher.

Example 4.5

Consider the following information:

Manufacturing rate of the predecessor facility in units per year, P = 1250
Demand rate of the successor facility in units per year, D = 750

Annual production-inventory carrying cost per unit, $I = ₹50$

Facility setup cost, $C = ₹75$

Annual backordering cost per unit, $K = ₹100$

Now:

Maximum inventory level, V can be obtained as:

$$\sqrt{\frac{2CKD(1 - D/P)}{I(I + K)}}$$

$$= 24.49 \text{ units}$$

In case where the facility setup cost (C) is increased to ₹90, then the revised V value is obtained as:

$$V_1 = \sqrt{\frac{2C_1KD(1 - D/P)}{I(I + K)}}$$

$$= 26.83$$

For a generalized approach:

$$V_1 - V = \sqrt{\frac{2C_1KD(1 - D/P)}{I(I + K)}} - \sqrt{\frac{2CKD(1 - D/P)}{I(I + K)}}$$

Or:

$$V_1 - V = \sqrt{\frac{2CKD(1 - D/P)}{I(I + K)}} \left[\sqrt{\frac{C_1}{C}} - 1 \right]$$

And:

%e increase in V,

$$= \left[\sqrt{\frac{C_1}{C}} - 1 \right] * 100$$

$$= \left[\sqrt{\left(1 + \frac{W}{100} \right)} - 1 \right] * 100$$

where $W = \%$ increase in C.

Similarly, with the reduction in setup cost:

$$V - V_1 = \sqrt{\frac{2CKD(1 - D/P)}{I(I + K)}} \left[1 - \sqrt{\frac{C_1}{C}} \right]$$

And:

% reduction in V,

$$= \left[1 - \sqrt{\frac{C_1}{C}}\right] * 100$$

$$= \left[1 - \sqrt{\left(1 - \frac{W}{100}\right)}\right] * 100$$

where W = % reduction in C.

4.2   ENTREPRENEURIAL APPLICATION

If any operational factor changes, it affects the maximum inventory level and eventually the space requirement. In the context of manufacturing entrepreneurship, the generalized approaches along with the following examples would be of interest.

Example 4.6

Because of an increase in K, the V value also increases. In order to bring back the V value to an original level, the suitable options may include:

a. Reduction in C
b. Reduction in P
c. Simultaneous reduction in C and P

For the simultaneous reduction, let:
Reduced facility setup cost,

$$C_1 = C\left(1 - \frac{Z}{100}\right)$$

Reduced manufacturing rate,

$$P_1 = P\left(1 - \frac{S}{100}\right)$$

where Z and S are percentage reduction in C and P, respectively.

Also:

$$K_1 = K\left(1 + \frac{W}{100}\right)$$

where W is the percentage increase in K.

Now:

$$\sqrt{\frac{2C_1 K_1 D(1 - D/P_1)}{I(I + K_1)}} = \sqrt{\frac{2CKD(1 - D/P)}{I(I + K)}}$$

Or:

$$\frac{C_1 K_1 (1 - D / P_1)}{(I + K_1)} = \frac{CK(1 - D / P)}{(I + K)}$$

Or:

$$1 - \frac{D}{P_1} = \frac{CK(1 - D / P)(I + K_1)}{C_1 K_1 (I + K)}$$

Or:

$$1 - \frac{D}{P_1} = \frac{(1 - D / P)(I + K_1)}{(1 - Z / 100)(1 + W / 100)(I + K)}$$

Or:

$$\frac{D}{P_1} = \frac{(1 - Z / 100)(1 + W / 100)(I + K) - (1 - D / P)\{I + K(1 + W / 100)\}}{(1 - Z / 100)(1 + W / 100)(I + K)}$$

Or:

$$\frac{P_1}{D} = \frac{(1 - Z / 100)(1 + W / 100)(I + K)}{(1 - Z / 100)(1 + W / 100)(I + K) - (1 - D / P)\{I + K(1 + W / 100)\}}$$

Or:

$$1 - \frac{S}{100} = \frac{(D / P)(1 - Z / 100)(1 + W / 100)(I + K)}{(1 - Z / 100)(1 + W / 100)(I + K) - (1 - D / P)\{I + K(1 + W / 100)\}}$$

Or:

$$S = 100 \left[1 - \frac{(D / P)(1 - Z / 100)(1 + W / 100)(I + K)}{(1 - Z / 100)(1 + W / 100)(I + K) - (1 - D / P)\{I + K(1 + W / 100)\}} \right] \quad (4.11)$$

For option (a), $S = 0$, and substituting this:

$$(1 - Z / 100)(1 + W / 100)(I + K) - (1 - D / P)\{I + K(1 + W / 100)\} =$$
$$(D / P)(1 - Z / 100)(1 + W / 100)(I + K)$$

Or:

$$(1 - Z / 100)(1 + W / 100)(I + K)(1 - D / P) = (1 - D / P)\{I + K(1 + W / 100)\}$$

Or:

$$1 - \frac{Z}{100} = \frac{(1 - D / P)\{I + K(1 + W / 100)\}}{(1 + W / 100)(I + K)(1 - D / P)}$$

Or:

$$Z = 100 \left[1 - \frac{I + K(1 + W / 100)}{(1 + W / 100)(I + K)} \right]$$

TABLE 4.7
Values of Z (Facility Setup Cost) Corresponding to W(K)

S. No.	W	$Z = 100\left[1 - \dfrac{I + K(1 + W/100)}{(1 + W/100)(I + K)}\right]$
1	10	3.03
2	20	5.56
3	30	7.69
4	40	9.52
5	50	11.11

For the given values as:

Manufacturing rate in units per year, P = 1250
Demand rate in units per year, D = 750
Annual production-inventory carrying cost per unit, I = ₹50
Facility setup cost, C = ₹75
Annual backordering cost per unit, K = ₹100

And with the use of relevant data, values of Z corresponding to W are provided in Table 4.7. This is in reference to the percentage reduction in facility setup cost (Z) in order to deal with the percentage increase in backordering cost, K, i.e., the value of W. Also, this is concerning the first option, i.e., the option (a) reduction in the facility setup cost alone in the manufacturing environment. If the backordering cost increases in the industrial/business environment, the entrepreneurial manager has one of the options in terms of the reduction in facility setup cost along with the appropriate quantification, helpful in the analysis of the situation encountered because of one or other problem being faced. Use of the option depends on the feasibility/applicability including the extent of reduction.

For option (b), i.e., the reduction in manufacturing rate (P):

Z = 0; Substituting in Eq. (4.11):

$$S = 100\left[1 - \frac{(D/P)(1 + W/100)(I + K)}{(1 + W/100)(I + K) - (1 - D/P)\{I + K(1 + W/100)\}}\right]$$

Table 4.8 shows the values of S corresponding to W for the given data.

For the option (c), i.e., the simultaneous reduction in C and P, various combinations of Z and S are provided in Table 4.9 for the value of W = 20. With the use of Eq. (4.11), such combinations are evaluated for the purpose of providing a wider choice in terms of factors and their values. Depending on the ease of implementation, the appropriate combination might be used, where the ease of implementation also relates to the level of quantitative variation in the operational factors among other aspects.

TABLE 4.8

Values of S (Manufacturing Rate) Corresponding to W(K)

S. No.	W	$S = 100\left[1 - \dfrac{(D/P)(1+W/100)(I+K)}{(1+W/100)(I+K)-(1-D/P)\{I+K(1+W/100)\}}\right]$
1	10	1.98
2	20	3.57
3	30	4.88

TABLE 4.9

Combination of Z and S for W = 20

S. No.	Z	$S = 100\left[1 - \dfrac{(D/P)(1-Z/100)(1+W/100)(I+K)}{(1-Z/100)(1+W/100)(I+K)-(1-D/P)\{I+K(1+W/100)\}}\right]$
1	1	2.98
2	2	2.36
3	3	1.73
4	4	1.07

Example 4.7

With increase in the manufacturing rate, the V value increases. For the goal of similar V value, the following suitable remedial measures may be available:

  i. Reduction in C
 ii. Reduction in K
iii. Simultaneous reduction in C and K

For the simultaneous reduction, let:

$$C_1 = C\left(1 - \frac{Z}{100}\right)$$

$$K_1 = K\left(1 - \frac{S}{100}\right)$$

where Z and S are % reduction in C and K, respectively.

Also:

$$P_1 = P\left(1 + \frac{W}{100}\right)$$

Now:

$$\sqrt{\frac{2C_1K_1D(1-D/P_1)}{I(I+K_1)}} = \sqrt{\frac{2CKD(1-D/P)}{I(I+K)}}$$

Or:

$$\frac{C_1 K_1(1 - D / P_1)}{(I + K_1)} = \frac{CK(1 - D / P)}{(I + K)}$$

Or:

$$\frac{(I + K_1)}{K_1} = \frac{(I + K)C_1(1 - D / P_1)}{CK(1 - D / P)}$$

Or:

$$\frac{I}{K_1} = \frac{(I + K)C_1(1 - D / P_1) - CK(1 - D / P)}{CK(1 - D / P)}$$

Or:

$$\frac{I}{K_1} = \frac{(I + K)(1 - Z / 100)(1 - D / P_1) - K(1 - D / P)}{K(1 - D / P)}$$

Or:

$$\frac{K_1}{I} = \frac{K(1 - D / P)}{(I + K)(1 - Z / 100)(1 - D / P_1) - K(1 - D / P)}$$

Or:

$$1 - \frac{S}{100} = \frac{I(1 - D / P)}{(I + K)(1 - Z / 100)\{1 - D / P(1 + W / 100)\} - K(1 - D / P)}$$

Or:

$$\frac{S}{100} = 1 - \frac{I(1 - D / P)}{(I + K)(1 - Z / 100)\{1 - D / P(1 + W / 100)\} - K(1 - D / P)}$$

Or:

$$S = 100\left[1 - \frac{I(1 - D / P)}{(I + K)(1 - Z / 100)\{1 - D / P(1 + W / 100)\} - K(1 - D / P)}\right] \quad (4.12)$$

For the first option:

$S = 0$, and:

$$(I + K)(1 - Z / 100)\{1 - D / P(1 + W / 100)\} - K(1 - D / P) = I(1 - D / P)$$

Or:

$$1 - \frac{Z}{100} = \frac{(I + K)(1 - D / P)}{(I + K)\{1 - D / P(1 + W / 100)\}}$$

Or:

$$\frac{Z}{100} = 1 - \frac{(1 - D / P)}{1 - D / \{P(1 + W / 100)\}}$$

Or:

$$Z = 100\left[1 - \frac{(1 - D / P)}{1 - D / \{P(1 + W / 100)\}}\right]$$

TABLE 4.10

Values of Z (Facility Setup Cost) Corresponding to W(P)

S. No.	W	$Z = 100\left[1 - \dfrac{(1 - D/P)}{1 - D/\{P(1 + W/100)\}}\right]$
1	5	6.67
2	10	12.00
3	15	16.36
4	20	20.00
5	25	23.08

For the relevant values such as:

Manufacturing rate in units per year, P = 1250
Demand rate in units per year, D = 750

Values of Z are given in Table 4.10 corresponding to W pertaining to the manufacturing rate.

For the second option:

Z = 0, and from Eq. (4.12),

$$S = 100\left[1 - \frac{I(1 - D/P)}{(I + K)\{1 - D/P(1 + W/100)\} - K(1 - D/P)}\right]$$

For the relevant values such as:

Manufacturing rate in units per year, P = 1250
Demand rate in units per year, D = 750
Annual production-inventory carrying cost per unit, I = ₹50
Annual backordering cost per unit, K = ₹100

Values of S are given in Table 4.11 corresponding to W pertaining to the manufacturing rate.

TABLE 4.11

Values of S (Backordering Cost) Corresponding to W(P)

S. No.	W	$S = 100\left[1 - \dfrac{I(1 - D/P)}{(I + K)\{1 - D/P(1 + W/100)\} - K(1 - D/P)}\right]$
1	5	17.65
2	10	29.03
3	15	36.99
4	20	42.86
5	25	47.37

TABLE 4.12
Combination of Z and S for W = 10

S. No.	Z	$S = 100\left[1 - \dfrac{I(1-D/P)}{(I+K)(1-Z/100)\{1-D/P(1+W/100)\} - K(1-D/P)}\right]$
1	2	25.42
2	4	21.43
3	6	16.98
4	8	12.00

For the third option, various combinations of Z and S are provided in Table 4.12 with the use of Eq. (4.12) for W = 10.

Example 4.8

With a reduction in the holding cost, V value increases. The suitable options may include:

a. Reduction in K
b. Reduction in P
c. Reduction in C
d. Simultaneous reduction of K and P
e. Simultaneous reduction of P and C
f. Simultaneous reduction of C and K
g. Simultaneous reduction of all the three factors, i.e., K, P, and C

For the simultaneous reduction of all the three factors, let:

$$K_1 = K\left(1 - \frac{R}{100}\right)$$

$$P_1 = P\left(1 - \frac{Z}{100}\right)$$

$$C_1 = C\left(1 - \frac{S}{100}\right)$$

where R, Z, and S are the % reduction in K, P, and C, respectively.
And also:

$$I_1 = I\left(1 - \frac{W}{100}\right)$$

where W is the % reduction in I.
Now:

$$\sqrt{\frac{2C_1K_1D(1-D/P_1)}{I_1(I_1+K_1)}} = \sqrt{\frac{2CKD(1-D/P)}{I(I+K)}}$$

Or:

$$\frac{C_1 K_1(1 - D/P_1)}{I_1(I_1 + K_1)} = \frac{CK(1 - D/P)}{(I + K)}$$

Or:

$$C_1 = \frac{I_1(I_1 + K_1)CK(1 - D/P)}{I(I + K)K_1(1 - D/P_1)}$$

Or:

$$1 - \frac{S}{100} = \frac{(1 - W/100)\{I(1 - W/100) + K(1 - R/100)\}(1 - D/P)}{(I + K)(1 - R/100)\{1 - D/P(1 - Z/100)\}}$$

Or:

$$S = 100\left[1 - \frac{(1 - W/100)\{I(1 - W/100) + K(1 - R/100)\}(1 - D/P)}{(I + K)(1 - R/100)\{1 - D/P(1 - Z/100)\}}\right] \quad (4.13)$$

Now various options may be analyzed.

a. Reduction in K:

$$Z = 0, \ S = 0$$

From Eq. (4.13),

$$(1 - W/100)\{I(1 - W/100) + K(1 - R/100)\}(1 - D/P) = (I + K)(1 - R/100)(1 - D/P)$$

Or:

$$I(1 - W/100)^2 + K(1 - R/100)(1 - W/100) = (I + K) - (I + K)(R/100)$$

Or:

$$(I + K)(R/100) + K(1 - R/100)(1 - W/100) = (I + K) - I(1 - W/100)^2$$

Or:

$$(I + K)(R/100) + K(1 - W/100) - K(R/100)(1 - W/100) = (I + K) - I(1 - W/100)^2$$

Or:

$$(R/100)\{I + K - K(1 - W/100)\} = (I + K) - I(1 - W/100)^2 - K(1 - W/100)$$

Or:

$$(R/100)\{I + K(W/100)\} = I - I(1 - W/100)^2 + K(W/100)$$

Or:

$$R = 100\left[\frac{I - I(1 - W/100)^2 + K(W/100)}{I + K(W/100)}\right]$$

For the given input parameters as:

Annual production-inventory carrying cost per unit, I = ₹50
Annual backordering cost per unit, K = ₹100

TABLE 4.13
Values of R (Backordering Cost) Corresponding to W(I)

S. No.	W	$R = 100\left[\dfrac{I - I(1 - W/100)^2 + K(W/100)}{I + K(W/100)}\right]$
1	5	17.95
2	10	32.50
3	15	44.42
4	20	54.29
5	25	62.50

Values of R have been given in Table 4.13 corresponding to W related to the inventory carrying cost.

b. Reduction in P:

$$R = 0, S = 0$$

From Eq. (4.13),

$$(1 - W/100)\{I(1 - W/100) + K\}(1 - D/P) = (I + K)\{1 - D/P(1 - Z/100)\}$$

Or:

$$1 - \frac{D}{P(1 - Z/100)} = \frac{(1 - W/100)(1 - D/P)\{I(1 - W/100) + K\}}{(I + K)}$$

Or:

$$\frac{D}{P(1 - Z/100)} = \frac{(I + K) - (1 - W/100)(1 - D/P)\{I(1 - W/100) + K\}}{(I + K)}$$

Or:

$$\frac{P(1 - Z/100)}{D} = \frac{(I + K)}{(I + K) - (1 - W/100)(1 - D/P)\{I(1 - W/100) + K\}}$$

Or:

$$1 - \frac{Z}{100} = \frac{(D/P)(I + K)}{(I + K) - (1 - W/100)(1 - D/P)\{I(1 - W/100) + K\}}$$

Or:

$$Z = 100\left[1 - \frac{(D/P)(I + K)}{(I + K) - (1 - W/100)(1 - D/P)\{I(1 - W/100) + K\}}\right]$$

For the given parameters:

Manufacturing rate in units per year, P = 1250
Demand rate in units per year, D = 750
Annual production-inventory carrying cost per unit, I = ₹50
Annual backordering cost per unit, K = ₹100

TABLE 4.14

Values of Z (Manufacturing Rate) Corresponding to W(I)

S. No.	W	$Z = 100\left[1 - \dfrac{(D/P)(I+K)}{(I+K)-(1-W/100)(1-D/P)\{I(1-W/100)+K\}}\right]$
1	5	4.20
2	10	7.97
3	15	11.37
4	20	14.45
5	25	17.24

Values of Z are provided in Table 4.14 along with W related to inventory carrying cost. The Z values are lower than the W values.

c. Reduction in C:

$$R = 0, Z = 0$$

From Eq. (4.13),

$$S = 100\left[1 - \frac{(1-W/100)\{I(1-W/100)+K\}(1-D/P)}{(I+K)\{1-D/P\}}\right]$$

Or:

$$S = 100\left[1 - \frac{(1-W/100)\{I(1-W/100)+K\}}{(I+K)}\right]$$

For the available parameters, values of S are provided in Table 4.15 along with W related to inventory carrying cost. The S values are higher than the related W values used in the numerical example.

TABLE 4.15

Values of S (Facility Setup Cost) Corresponding to W (I)

S. No.	W	$S = 100\left[1 - \dfrac{(1-W/100)\{I(1-W/100)+K\}}{(I+K)}\right]$
1	5	6.58
2	10	13.00
3	15	19.25
4	20	25.33
5	25	31.25

d. Simultaneous reduction of K and P:

$S = 0$, and from Eq. (4.13):

$$(I+K)(1-R/100)\{1-D/P(1-Z/100)\} =$$
$$(1-W/100)\{I(1-W/100)+K(1-R/100)\}(1-D/P)$$

Or:

$$1-\frac{D}{P(1-Z/100)} = \frac{(1-W/100)(1-D/P)\{I(1-W/100)+K(1-R/100)\}}{(I+K)(1-R/100)}$$

Or:

$$\frac{D}{P(1-Z/100)} = \frac{(I+K)(1-R/100)-(1-W/100)(1-D/P)\{I(1-W/100)+K(1-R/100)\}}{(I+K)(1-R/100)}$$

Or:

$$1-\frac{Z}{100} = \frac{(D/P)(I+K)(1-R/100)}{(I+K)(1-R/100)-(1-W/100)(1-D/P)\{I(1-W/100)+K(1-R/100)\}}$$

Or:

$$Z = 100\left[1-\frac{(D/P)(I+K)(1-R/100)}{(I+K)(1-R/100)-(1-W/100)(1-D/P)\{I(1-W/100)+K(1-R/100)\}}\right]$$

Various combinations of R and Z have been provided in Table 4.16 for W = 15 and the available parameters.

e. Simultaneous reduction of P and C:

$R = 0$, and from Eq. (4.13):

$$S = 100\left[1-\frac{(1-W/100)\{I(1-W/100)+K\}(1-D/P)}{(I+K)\{1-D/P(1-Z/100)\}}\right]$$

Various combinations of Z and S have been provided in Table 4.17 for W = 15.

TABLE 4.16
Combination of R and Z for W = 15

S. No.	R	$Z = 100\left[1-\dfrac{(D/P)(I+K)(1-R/100)}{(I+K)(1-R/100)-(0.85)(1-D/P)\{(0.85)I+K(1-R/100)\}}\right]$
1	5	10.70
2	10	9.95
3	15	9.09
4	20	8.10
5	25	6.96

TABLE 4.17

Combination of Z and S for W = 15

S. No.	Z	$S = 100\left[1 - \dfrac{(1-W/100)\{I(1-W/100)+K\}(1-D/P)}{(I+K)\{1-D/P(1-Z/100)\}}\right]$
1	1	18.01
2	3	15.32
3	5	12.33
4	7	8.97
5	9	5.18

f. Simultaneous reduction of C and K:

Z = 0, and from Eq. (4.13):

$$S = 100\left[1 - \frac{(1-W/100)\{I(1-W/100)+K(1-R/100)\}(1-D/P)}{(I+K)(1-R/100)\{1-D/P\}}\right]$$

Or:

$$S = 100\left[1 - \frac{(1-W/100)\{I(1-W/100)+K(1-R/100)\}}{(I+K)(1-R/100)}\right]$$

Various combinations of R and S have been provided in Table 4.18 for W = 15 and the available parameters.

g. Simultaneous reduction of all the three factors, i.e., K, P, and C:

Refer to Table 4.18. For W = 15, the desired objective of similar V can be achieved, for example, by:

R = 15
S = 15

For instance, if S cannot be varied by more than 10, then consider:

S = 10

TABLE 4.18

Combination of R and S for W = 15

S. No.	R	$S = 100\left[1 - \dfrac{(1-W/100)\{I(1-W/100)+K(1-R/100)\}}{(I+K)(1-R/100)}\right]$
1	5	17.98
2	10	16.57
3	15	15.00
4	20	13.23
5	25	11.22

Now, from Eq. (4.13):

$$S = 100\left[1 - \frac{(1-W/100)\{I(1-W/100)+K(1-R/100)\}(1-D/P)}{(I+K)(1-R/100)\{1-D/P(1-Z/100)\}}\right]$$

For the given parameters:

Manufacturing rate in units per year, P = 1250
Demand rate in units per year, D = 750
Annual production-inventory carrying cost per unit, I = ₹50
Annual backordering cost per unit, K = ₹100

And also, with:

W = 15
R = 15
S = 10

Value of Z from the Eq. (4.13), for the desired objective, can be evaluated as:

Z = 3.57

Thus, the simultaneous reduction of all the three factors, i.e., K, P, and C may be implemented by corresponding values of R, Z, and S as follows:

R = 15
Z = 3.57
S = 10

Example 4.9

With an increase in value of C, the V value increases. To deal with this, the following remedial measures may be considered:

  i. Reduction in K
 ii. Reduction in P
iii. Simultaneous reduction in K and P

For the simultaneous reduction, let:

$$K_1 = K\left(1 - \frac{Z}{100}\right)$$

$$P_1 = P\left(1 - \frac{S}{100}\right)$$

And:

$$C_1 = C\left(1 + \frac{W}{100}\right)$$

Now:

$$\sqrt{\frac{2C_1K_1D(1-D/P_1)}{I(I+K_1)}} = \sqrt{\frac{2CKD(1-D/P)}{I(I+K)}}$$

Or:

$$\frac{C_1K_1(1-D/P_1)}{(I+K_1)} = \frac{CK(1-D/P)}{(I+K)}$$

Or:

$$1-\frac{D}{P_1} = \frac{CK(1-D/P)(I+K_1)}{(I+K)C_1K_1}$$

Or:

$$\frac{D}{P_1} = \frac{(I+K)C_1K_1 - CK(1-D/P)(I+K_1)}{(I+K)C_1K_1}$$

Or:

$$\frac{P_1}{D} = \frac{(I+K)C_1K_1}{(I+K)C_1K_1 - CK(1-D/P)(I+K_1)}$$

Or:

$$1-\frac{S}{100} = \frac{(D/P)(I+K)(1+W/100)(1-Z/100)}{(I+K)(1+W/100)(1-Z/100)-(1-D/P)(I+K_1)}$$

Or:

$$S = 100\left[1-\frac{(D/P)(I+K)(1+W/100)(1-Z/100)}{(I+K)(1+W/100)(1-Z/100)-(1-D/P)\{I+K(1-Z/100)\}}\right] \quad (4.14)$$

For the first option, i.e., the reduction in K:

S = 0, and from Eq. (4.14),

$$(I+K)(1+W/100)(1-Z/100)-(1-D/P)\{I+K(1-Z/100)\} =$$
$$(D/P)(I+K)(1+W/100)(1-Z/100)$$

Or:

$$(I+K)(1+W/100)(1-Z/100)(1-D/P)-(1-D/P)K(1-Z/100) = I(1-D/P)$$

Or:

$$1-\frac{Z}{100} = \frac{I}{(I+K)(1+W/100)-K}$$

Or:

$$Z = 100\left[1-\frac{I}{(I+K)(1+W/100)-K}\right]$$

For the given parameters:

Manufacturing rate in units per year, P = 1250
Demand rate in units per year, D = 750
Annual production-inventory carrying cost per unit, I = ₹50

TABLE 4.19
Values of Z (Backordering Cost) Corresponding to W(C)

S. No.	W	$Z = 100\left[1 - \dfrac{I}{(I + K)(1 + W / 100) - K}\right]$
1	5	13.04
2	10	23.08
3	15	31.03
4	20	37.50
5	25	42.86

Annual backordering cost per unit, K = ₹100
And with the use of relevant factors

Values of Z are provided in Table 4.19 along with W related to facility setup cost. For the second option, i.e., the reduction in P:

Z = 0, and from Eq. (4.14),

$$S = 100\left[1 - \frac{(D / P)(I + K)(1 + W / 100)}{(I + K)(1 + W / 100) - (1 - D / P)(I + K)}\right]$$

Or:

$$S = 100\left[1 - \frac{(D / P)(1 + W / 100)}{(1 + W / 100) - (1 - D / P)}\right]$$

Values of S are provided in Table 4.20 corresponding to W related to the facility setup cost.
For the third option, i.e., the simultaneous reduction in K and P:
From Eq. (4.14),

$$S = 100\left[1 - \frac{(D / P)(I + K)(1 + W / 100)(1 - Z / 100)}{(I + K)(1 + W / 100)(1 - Z / 100) - (1 - D / P)\{I + K(1 - Z / 100)\}}\right]$$

TABLE 4.20
Values of S (Manufacturing Rate) Corresponding to W (C)

S. No.	W	$S = 100\left[1 - \dfrac{(D / P)(1 + W / 100)}{(1 + W / 100) - (1 - D / P)}\right]$
1	5	3.08
2	10	5.71
3	15	8.00
4	20	10.00
5	25	11.76

TABLE 4.21

Combination of Z and S for W = 20

S. No.	Z	$S = 100\left[1 - \dfrac{(D/P)(I+K)(1+W/100)(1-Z/100)}{(I+K)(1+W/100)(1-Z/100)-(1-D/P)\{I+K(1-Z/100)\}}\right]$
1	3	9.53
2	6	9.03
3	9	8.49
4	12	7.91
5	15	7.27

And: For W = 20;

Different combinations of Z and S have been provided in Table 4.21 considering the available relevant factors.

Example 4.10

Although demand can be an external factor, its variation has an influence on the space requirement. If demand rate decreases, V value may increase. For bringing the V value to an original level, the following measures may be adopted:

a. Reduction in P
b. Reduction in C
c. Reduction in K
d. Simultaneous reduction in P and C
e. Simultaneous reduction in C and K
f. Simultaneous reduction in K and P
g. Simultaneous reduction in all the three factors, i.e., P, C, and K.

For the simultaneous reduction in all the three factors, let:

$$P_1 = P\left(1 - \frac{R}{100}\right)$$

$$C_1 = C\left(1 - \frac{Z}{100}\right)$$

$$K_1 = K\left(1 - \frac{S}{100}\right)$$

And also:

$$D_1 = D\left(1 - \frac{W}{100}\right)$$

Now:

$$\sqrt{\frac{2C_1 K_1 D_1(1 - D_1/P_1)}{I(I + K_1)}} = \sqrt{\frac{2CKD(1 - D/P)}{I(I + K)}}$$

Or:

$$\frac{C_1}{C} = \frac{KD(1-D/P)(I+K_1)}{K_1D_1(1-D_1/P_1)(I+K)}$$

Or:

$$1 - \frac{Z}{100} = \frac{(1-D/P)\{I + K(1-S/100)\}}{(I+K)(1-S/100)(1-W/100)\{1-(D/P)(1-W/100)/(1-R/100)\}}$$

Or:

$$Z = 100\left[1 - \frac{(1-D/P)\{I + K(1-S/100)\}}{(I+K)(1-S/100)(1-W/100)\{1-(D/P)(1-W/100)/(1-R/100)\}}\right]$$

$$(4.15)$$

Now each remedial measure can be analyzed.

a. Reduction in P

$Z = 0$, $S = 0$, and from Eq. (4.15):

$$(I+K)(1-W/100)\{1-(D/P)(1-W/100)/(1-R/100)\} = (1-D/P)(I+K)$$

Or:

$$1 - \frac{(D/P)(1-W/100)}{(1-R/100)} = \frac{(1-D/P)}{(1-W/100)}$$

Or:

$$\frac{(D/P)(1-W/100)}{(1-R/100)} = \frac{(D/P)-(W/100)}{(1-W/100)}$$

Or:

$$\frac{(1-R/100)}{(D/P)(1-W/100)} = \frac{(1-W/100)}{(D/P)-(W/100)}$$

Or:

$$1 - \frac{R}{100} = \frac{(D/P)(1-W/100)^2}{(D/P)-(W/100)}$$

Or:

$$R = 100\left[1 - \frac{(D/P)(1-W/100)^2}{(D/P)-(W/100)}\right]$$

With the use of relevant factors such as:

Manufacturing rate in units per year, $P = 1250$
Demand rate in units per year, $D = 750$

Values of R have been given in Table 4.22 corresponding to W related to the demand rate. These values of R, concerning the rate of manufacture, are much lower with reference to the W.

TABLE 4.22

Values of R (Manufacturing Rate) Corresponding to W(D)

S. No.	W	$R = 100\left[1 - \dfrac{(D/P)(1-W/100)^2}{(D/P)-(W/100)}\right]$
1	5	1.55
2	10	2.80
3	15	3.67
4	20	4.00

b. Reduction in C

R = 0, S = 0, and from Eq. (4.15):

$$Z = 100\left[1 - \frac{(1-D/P)(I+K)}{(I+K)(1-W/100)\{1-(D/P)(1-W/100)\}}\right]$$

Or:

$$Z = 100\left[1 - \frac{(1-D/P)}{(1-W/100)\{1-(D/P)(1-W/100)\}}\right]$$

With the related factors such as:

Annual production-inventory carrying cost per unit, I = ₹50
Annual backordering cost per unit, K = ₹100
Manufacturing rate in units per year, P = 1250
Demand rate in units per year, D = 750

Values of Z have been evaluated in Table 4.23 corresponding to W related to the demand rate.

c. Reduction in K

R = 0, Z = 0, and from Eq. (4.15):

$$(I+K)(1-S/100)(1-W/100)\{1-(D/P)(1-W/100)\} = (1-D/P)\{I+K(1-S/100)\}$$

TABLE 4.23

Values of Z (Facility Setup Cost) Corresponding to W(D)

S. No.	W	$Z = 100\left[1 - \dfrac{(1-D/P)}{(1-W/100)\{1-(D/P)(1-W/100)\}}\right]$
1	5	2.08
2	10	3.38
3	15	3.96
4	20	3.85

TABLE 4.24

Values of S (Backordering Cost) Corresponding to W(D)

S. No.	W	$S = 100\left[1 - \dfrac{I(1-D/P)}{(I+K)(1-W/100)\{1-(D/P)(1-W/100)\} - K(1-D/P)}\right]$
1	5	5.99
2	10	9.50
3	15	11.01
4	20	10.71

Or:

$$(I+K)(1-S/100)(1-W/100)\{1-(D/P)(1-W/100)\}$$
$$= I(1-D/P) + K(1-D/P)(1-S/100)$$

Or:

$$(1-S/100)\left[(I+K)(1-W/100)\{1-(D/P)(1-W/100)\} - K(1-D/P)\right] = I(1-D/P)$$

Or:

$$1 - \frac{S}{100} = \frac{I(1-D/P)}{(I+K)(1-W/100)\{1-(D/P)(1-W/100)\} - K(1-D/P)}$$

Or:

$$S = 100\left[1 - \frac{I(1-D/P)}{(I+K)(1-W/100)\{1-(D/P)(1-W/100)\} - K(1-D/P)}\right]$$

Values of S have been evaluated with the given factor values in Table 4.24 corresponding to W related to the demand rate.

d. Simultaneous reduction in P and C

S = 0, and from Eq. (4.15):

$$Z = 100\left[1 - \frac{(1-D/P)(I+K)}{(I+K)(1-W/100)\{1-(D/P)(1-W/100)/(1-R/100)\}}\right]$$

Or:

$$Z = 100\left[1 - \frac{(1-D/P)}{(1-W/100)\{1-(D/P)(1-W/100)/(1-R/100)\}}\right]$$

For W = 20, combinations of R and Z have been provided in Table 4.25 considering the given factor values. Suitable combination might be adopted on the basis of ease in implementation.

TABLE 4.25

Combination of R and Z for W = 20

S. No.	R	$Z = 100\left[1 - \dfrac{(1-D/P)}{(1-W/100)\{1-(D/P)(1-W/100)/(1-R/100)\}}\right]$
1	1	2.94
2	2	2.00
3	3	1.02

e. Simultaneous reduction in C and K

R = 0, and from Eq. (4.15):

$$Z = 100\left[1 - \frac{(1-D/P)\{I+K(1-S/100)\}}{(I+K)(1-S/100)(1-W/100)\{1-(D/P)(1-W/100)\}}\right]$$

For W = 15, combinations of S and Z have been provided in Table 4.26 considering the given factor values.

f. Simultaneous reduction in K and P

Z = 0, and from Eq. (4.15):

$$(I+K)(1-S/100)(1-W/100)\{1-(D/P)(1-W/100)/(1-R/100)\}$$
$$= (1-D/P)\{I+K(1-S/100)\}$$

Or:

$$1 - \frac{(D/P)(1-W/100)}{(1-R/100)} = \frac{(1-D/P)\{I+K(1-S/100)\}}{(I+K)(1-S/100)(1-W/100)}$$

Or:

$$\frac{(D/P)(1-W/100)}{(1-R/100)} = \frac{(I+K)(1-S/100)(1-W/100)-(1-D/P)\{I+K(1-S/100)\}}{(I+K)(1-S/100)(1-W/100)}$$

TABLE 4.26

Combination of S and Z for W = 15

S. No.	S	$Z = 100\left[1 - \dfrac{(1-D/P)\{I+K(1-S/100)\}}{(I+K)(1-S/100)(1-W/100)\{1-(D/P)(1-W/100)\}}\right]$
1	2	3.31
2	4	2.63
3	6	1.92
4	8	1.18
5	10	0.40

TABLE 4.27

Combination of S and R for W = 15

S. No.	S	$R = 100\left[1 - \dfrac{(D/P)(I+K)(1-S/100)(1-W/100)^2}{(I+K)(1-S/100)(1-W/100) - (1-D/P)\{I+K(1-S/100)\}}\right]$
1	1	2.40
2	2	1.10

Or:

$$\frac{(1-R/100)}{(D/P)(1-W/100)} = \frac{(I+K)(1-S/100)(1-W/100)}{(I+K)(1-S/100)(1-W/100) - (1-D/P)\{I+K(1-S/100)\}}$$

Or:

$$1 - \frac{R}{100} = \frac{(D/P)(I+K)(1-S/100)(1-W/100)^2}{(I+K)(1-S/100)(1-W/100) - (1-D/P)\{I+K(1-S/100)\}}$$

Or:

$$R = 100\left[1 - \frac{(D/P)(I+K)(1-S/100)(1-W/100)^2}{(I+K)(1-S/100)(1-W/100) - (1-D/P)\{I+K(1-S/100)\}}\right]$$

For W = 15, in order to illustrate, the two combinations of S and R have been provided in Table 4.27 considering the given factor values.

g. Simultaneous reduction in all the three factors, i.e., P, C, and K

In order to illustrate, refer Table 4.27. For W = 15:

S = 1.00
R = 2.40

In case where R cannot exceed 2.00, then all the three factors may be varied. Now, from Eq. (4.15):

$$Z = 100\left[1 - \frac{(1-D/P)\{I+K(1-S/100)\}}{(I+K)(1-S/100)(1-W/100)\{1-(D/P)(1-W/100)/(1-R/100)\}}\right]$$

Consider the following information:

Manufacturing rate in units per year, P = 1250
Demand rate in units per year, D = 750
Annual production-inventory carrying cost per unit, I = ₹50
Facility setup cost, C = ₹75
Annual backordering cost per unit, K = ₹100

With the use of relevant factors and also:

W = 15
R = 2
S = 1

The value of Z can be obtained from the Eq. (4.15) as follows:

Z = 1.55

4.3 QUALITY INCLUSION

The subsequent text discusses quality inclusion along with the backorders.

4.3.1 QUALITY INCLUSION ALONG WITH BACKORDERS

In order to find out the maximum inventory level during the manufacturing cycle with quality inclusion, refer to Figure 4.4, where:

y = Proportion of acceptable components in a manufacturing batch
V = Maximum production-inventory level concerning acceptable components

Now:
Manufacturing cycle time,

$$T = \frac{(V + J)}{(yP - D)} + \frac{(V + J)}{D}$$

Or:

$$T = (V + J)\left[\frac{D + yP - D)}{D(yP - D)}\right]$$

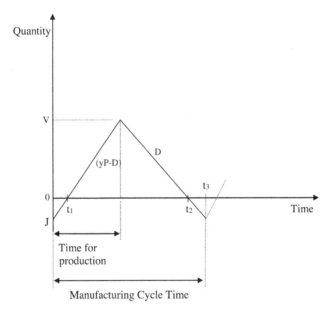

FIGURE 4.4   Quality inclusion along with backorders.

Or:

$$T = \frac{yP(V+J)}{D(yP-D)}$$

Or:

$$T = \frac{(V+J)}{D(1-D/yP)} \tag{4.16}$$

And, number of cycles in a year = 1/(cycle time, T)
Or:
  Number of cycles in a year,

$$= \frac{D(1-D/yP)}{(V+J)}$$

Since facility setup cost is incurred for each cycle, an annual facility setup cost is evaluated as the multiplication of such setup cost and the number of cycles in a year. Therefore:
  Annual facility setup cost,

$$= \frac{CD(1-D/yP)}{(V+J)} \tag{4.17}$$

where C = Facility setup cost.
  As the average inventory is,

$$\frac{V}{2}$$

And positive inventory exists in a cycle for the period:

$$\frac{V}{(yP-D)} + \frac{V}{D}$$

Annual production-inventory carrying cost can be evaluated as:

$$\frac{VI}{2}\left[\frac{V}{(yP-D)} + \frac{V}{D}\right] * \text{Number of cycles in a year}$$

where I = Annual production-inventory carrying cost per unit
Now:
  Annual production-inventory carrying cost,

$$= \frac{VI}{2}\left[\frac{V}{(yP-D)} + \frac{V}{D}\right]\frac{D(1-D/yP)}{(V+J)} \tag{4.18}$$

Similarly, period corresponding to backorders,

$$= \frac{J}{(yP-D)} + \frac{J}{D}$$

And, average backordering quantity,

$$= \frac{J}{2}$$

Also with, the number of cycles in a year,

$$= \frac{D(1 - D/yP)}{(V + J)}$$

The annual backordering cost,

$$= \frac{JK}{2} \left[\frac{J}{(yP - D)} + \frac{J}{D} \right] \frac{D(1 - D/yP)}{(V + J)} \tag{4.19}$$

where K = Annual backordering cost per unit.

4.3.2 TOTAL RELATED COST

Adding the expression (4.17), (4.18), and (4.19), the total related cost (E) is formulated as:

$$E = \frac{CD(1 - D/yP)}{(V + J)} + \frac{VI}{2} \left[\frac{V}{(yP - D)} + \frac{V}{D} \right] \frac{D(1 - D/yP)}{(V + J)}$$

$$+ \frac{JK}{2} \left[\frac{J}{(yP - D)} + \frac{J}{D} \right] \frac{D(1 - D/yP)}{(V + J)}$$

Or:

$$E = \frac{CD(1 - D/yP)}{(V + J)} + \frac{V^2 I}{2} \left[\frac{D + yP - D}{D(yP - D)} \right] \frac{D(1 - D/yP)}{(V + J)}$$

$$+ \frac{J^2 K}{2} \left[\frac{D + yP - D}{D(yP - D)} \right] \frac{D(1 - D/yP)}{(V + J)}$$

Or:

$$E = \frac{CD(1 - D/yP)}{(V + J)} + \frac{V^2 I}{2} \left[\frac{1}{D(1 - D/yP)} \right] \frac{D(1 - D/yP)}{(V + J)}$$

$$+ \frac{J^2 K}{2} \left[\frac{1}{D(1 - D/yP)} \right] \frac{D(1 - D/yP)}{(V + J)}$$

Or:

$$E = \frac{CD(1 - D/yP)}{(V + J)} + \frac{V^2 I}{2(V + J)} + \frac{J^2 K}{2(V + J)} \tag{4.20}$$

From Eq. (4.16):

$$V + J = TD(1 - D / yP) \tag{4.21}$$

Substituting in Eq. (4.20):

$$E = \frac{C}{T} + \frac{V^2 I}{2TD(1 - D / yP)} + \frac{J^2 K}{2TD(1 - D / yP)}$$

Using the value of J from Eq. (4.21),

$$E = \frac{C}{T} + \frac{V^2 I + K\left[TD(1 - D / yP) - V\right]^2}{2TD(1 - D / yP)}$$

Or:

$$E = \frac{C}{T} + \frac{V^2 I + K\left[T^2 D^2 (1 - D / yP)^2 - 2TDV(1 - D / yP) + V^2\right]}{2TD(1 - D / yP)}$$

Or:

$$E = \frac{C}{T} + \frac{V^2 (I + K) + KTD(1 - D / yP)\left[TD(1 - D / yP) - 2V\right]}{2TD(1 - D / yP)}$$

Or:

$$E = \frac{C}{T} + \frac{V^2 (I + K)}{2TD(1 - D / yP)} + \frac{KTD(1 - D / yP)}{2} - VK \tag{4.22}$$

With an objective of minimizing the total related cost, differentiating partially with respect to V and equating to zero,

$$\frac{V(I + K)}{TD(1 - D / yP)} = K$$

Or:

$$TD(1 - D / yP) = \frac{V(I + K)}{K}$$

Substituting in Eq. (4.22),

$$E = \frac{CKD(1 - D / yP)}{V(I + K)} + \frac{V^2 (I + K)K}{2V(I + K)} + \frac{KV(I + K)}{2K} - VK$$

Or:

$$E = \frac{CKD(1 - D / yP)}{V(I + K)} + \frac{VK}{2} + \frac{V(I + K)}{2} - VK$$

Or:

$$E = \frac{CKD(1 - D / yP)}{V(I + K)} + \frac{VI}{2} \tag{4.23}$$

Differentiating with respect to V and equating to zero,

$$\frac{I}{2} - \frac{CKD(1 - D/yP)}{V^2(I + K)} = 0$$

Or:

$$V^2 = \frac{2CKD(1 - D/yP)}{I(I + K)}$$

Or:

$$V^* = \sqrt{\frac{2CKD(1 - D/yP)}{I(I + K)}} \tag{4.24}$$

From the above expression:

$$\sqrt{\frac{2CKD(1 - D/yP)}{(I + K)}} = V^* \sqrt{I}$$

Substituting in Eq. (4.23),

$$E^* = \frac{IV^*}{2} + \frac{IV^*}{2}$$

Or:

$$E^* = IV^* \tag{4.25}$$

4.3.3 VARIATION OF INVENTORY WITH RESPECT TO THE QUALITY LEVEL

This is illustrated with the following example.

Example 4.11

Consider the following details:

Proportion of acceptable components in a manufacturing batch, y = 0.95
Annual backordering cost per unit, K = ₹100
Manufacturing rate in units per year, P = 1250
Demand rate in units per year, D = 750
Annual production-inventory carrying cost per unit, I = ₹50
Facility setup cost, C = ₹75

Now:
From the Eq. (4.24), the value of maximum inventory level, V can be obtained as follows:

$$V = \sqrt{\frac{2CKD(1 - D/yP)}{I(I + K)}}$$

$$= 23.51 \text{ units}$$

TABLE 4.28

Variation of V with respect to y

S. No.	y	V
1	0.95	23.51
2	0.90	22.36
3	0.85	21.00
4	0.8	19.36
5	0.75	17.32

And, from Eq. (4.25), the total related cost can be obtained as follows:

$$E = 50 \times 23.51$$

$$= ₹1175.41$$

Table 4.28 shows the variation of V with respect to y value.

Example 4.12

With increase in y, V value increases. For bringing the V value to an original level, an option can be the reduction in K. For this purpose, let:

$$y_1 = y\left(1 + \frac{W}{100}\right)$$

$$K_1 = K\left(1 - \frac{Z}{100}\right)$$

Now:

$$\sqrt{\frac{2CK_1D(1 - D/y_1P)}{I(I + K_1)}} = \sqrt{\frac{2CKD(1 - D/yP)}{I(I + K)}}$$

Or:

$$\frac{K_1(1 - D/y_1P)}{(I + K_1)} = \frac{K(1 - D/yP)}{(I + K)}$$

Or:

$$\frac{(I + K_1)}{K_1} = \frac{(I + K)(1 - D/y_1P)}{K(1 - D/yP)}$$

Or:

$$\frac{I}{K_1} = \frac{(I + K)(1 - D/y_1P) - K(1 - D/yP)}{K(1 - D/yP)}$$

Or:

$$\frac{K_1}{I} = \frac{K(1-D/yP)}{(I+K)\{1-(D/yP)/(1+W/100)\} - K(1-D/yP)}$$

Or:

$$1 - \frac{Z}{100} = \frac{I(1-D/yP)}{(I+K)\{1-(D/yP)/(1+W/100)\} - K(1-D/yP)}$$

Or:

$$Z = 100\left[1 - \frac{I(1-D/yP)}{(I+K)\{1-(D/yP)/(1+W/100)\} - K(1-D/yP)}\right]$$

With the previous example data, values of Z are obtained in Table 4.29 corresponding to W related to the y.
   Along with K, the other factors can also be considered, such as:

a. Reduction of K and P
b. Reduction of K and C
c. Simultaneous reduction of K, P, and C

For the simultaneous reduction of K, P, and C, let:

$$K_1 = K\left(1 - \frac{Z}{100}\right)$$

$$P_1 = P\left(1 - \frac{R}{100}\right)$$

$$C_1 = C\left(1 - \frac{S}{100}\right)$$

And also:

$$y_1 = y\left(1 + \frac{W}{100}\right)$$

TABLE 4.29

Values of Z (Backordering Cost) Corresponding to W(y)

S. No.	W	$Z = 100\left[1 - \dfrac{I(1-D/yP)}{(I+K)\{1-(D/yP)/(1+W/100)\} - K(1-D/yP)}\right]$
1	1	4.84
2	2	9.16
3	3	13.03
4	4	16.51

Now:

$$\sqrt{\frac{2C_1 K_1 D(1 - D / y_1 P_1)}{I(I + K_1)}} = \sqrt{\frac{2CKD(1 - D / yP)}{I(I + K)}}$$

Or:

$$\frac{C_1 K_1 (1 - D / y_1 P_1)}{(I + K_1)} = \frac{CK(1 - D / yP)}{(I + K)}$$

Or:

$$\frac{C_1}{C} = \frac{K(I + K_1)(1 - D / yP)}{K_1 (I + K)(1 - D / y_1 P_1)}$$

Or:

$$1 - \frac{S}{100} = \frac{\{I + K(1 - Z / 100)\}(1 - D / yP)}{(1 - Z / 100)(I + K)(1 - D / y_1 P_1)}$$

Or:

$$1 - \frac{S}{100} = \frac{y_1 P_1 \{I + K(1 - Z / 100)\}(1 - D / yP)}{(1 - Z / 100)(I + K)(y_1 P_1 - D)}$$

Or:

$$1 - \frac{S}{100} = \frac{(1 + W / 100)(1 - R / 100)\{I + K(1 - Z / 100)\}(yP - D)}{(1 - Z / 100)(I + K)\{yP(1 + W / 100)(1 - R / 100) - D\}}$$

Or:

$$S = 100 \left[1 - \frac{(1 + W / 100)(1 - R / 100)\{I + K(1 - Z / 100)\}(yP - D)}{(1 - Z / 100)(I + K)\{yP(1 + W / 100)(1 - R / 100) - D\}} \right] \qquad (4.26)$$

Now the options (a), (b), and (c) can be analyzed.

a. Reduction of K and P

$S = 0$, and from Eq. (4.26):

$$(1 + W / 100)(1 - R / 100)\{I + K(1 - Z / 100)\}(yP - D)$$
$$= (1 - Z / 100)(I + K)\{yP(1 + W / 100)(1 - R / 100) - D\}$$

Or:

$$D(1 - Z / 100)(I + K) = (1 - Z / 100)(I + K)yP(1 + W / 100)(1 - R / 100)$$
$$- (1 + W / 100)(1 - R / 100)\{I + K(1 - Z / 100)\}(yP - D)$$

Or:

$$1 - \frac{R}{100} = \frac{D(1 - Z / 100)(I + K)}{(1 - Z / 100)(I + K)yP(1 + W / 100) - (1 + W / 100)\{I + K(1 - Z / 100)\}(yP - D)}$$

Or:

$$1 - \frac{R}{100} = \frac{D(1 - Z / 100)(I + K)}{(1 + W / 100)\left[yP(I + K)(1 - Z / 100) - \{I + K - (KZ / 100)\}(yP - D) \right]}$$

Or:

$$1 - \frac{R}{100} = \frac{D(1 - Z/100)(I + K)}{(1 + W/100)\left[\begin{array}{l} yP(I + K) - yP(I + K)(Z/100) - \\ \{yP(I + K) - yP(KZ/100) - D(I + K) + (DKZ/100)\} \end{array}\right]}$$

Or:

$$1 - \frac{R}{100} =$$

$$\frac{D(1 - Z/100)(I + K)}{(1 + W/100)\left[yP(KZ/100) + D(I + K) - (DKZ/100) - yPI(Z/100) - (yPK)(Z/100)\right]}$$

Or:

$$1 - \frac{R}{100} = \frac{D(1 - Z/100)(I + K)}{(1 + W/100)\left[D(I + K) - (DKZ/100) - yPI(Z/100)\right]}$$

Or:

$$\frac{R}{100} = 1 - \frac{D(1 - Z/100)(I + K)}{(1 + W/100)\{D(I + K) - (Z/100)(DK + yPI)\}}$$

Or:

$$R = 100\left[1 - \frac{D(1 - Z/100)(I + K)}{(1 + W/100)\{D(I + K) - (Z/100)(DK + yPI)\}}\right]$$

With the following data:

Proportion of acceptable components in a manufacturing batch, y = 0.95
Annual backordering cost per unit, K = ₹100
Manufacturing rate in units per year, P = 1250
Demand rate in units per year, D = 750
Annual production-inventory carrying cost per unit, I = ₹50
Facility setup cost, C = ₹75

For W = 4, combinations of Z and R have been provided in Table 4.30.

TABLE 4.30
Combination of Z and R for W = 4

S. No.	Z	$R = 100\left[1 - \dfrac{D(1 - Z/100)(I + K)}{(1 + W/100)\{D(I + K) - (Z/100)(DK + yPI)\}}\right]$
1	2	3.46
2	4	3.06
3	6	2.64
4	8	2.19
5	10	1.72

TABLE 4.31
Combination of Z and S for W = 4

S. No.	Z	$S = 100\left[1 - \dfrac{(1+W/100)\{I + K(1 - Z/100)\}(yP - D)}{(1 - Z/100)(I + K)\{yP(1 + W/100) - D\}}\right]$
1	2	5.55
2	4	4.88
3	6	4.19
4	8	3.47
5	10	2.71

b. Reduction of K and C

$R = 0$, and from Eq. (4.26):

$$S = 100\left[1 - \frac{(1+W/100)\{I + K(1 - Z/100)\}(yP - D)}{(1 - Z/100)(I + K)\{yP(1 + W/100) - D\}}\right]$$

For W = 4, combinations of Z and S have been provided in Table 4.31 considering the given data.

c. Simultaneous reduction of K, P, and C

Refer Table 4.9; for $Z = 2$, $R = 3.46$

For instance, if R cannot be reduced by more than 2, then:

For W = 4
Z = 2
R = 2

And S can be obtained with the use of Eq. (4.26) as follows:

$$S = 100\left[1 - \frac{(1+W/100)(1 - R/100)\{I + K(1 - Z/100)\}(yP - D)}{(1 - Z/100)(I + K)\{yP(1 + W/100)(1 - R/100) - D\}}\right]$$

$$= 2.47$$

As illustrated in this example, the simultaneous reduction of all the three factors can be implemented in case where it is needed such as:

For W = 4; Z = 2; R = 2; and S = 2.47,

where Z, R, and S correspond to percentage reduction in backordering cost (K), manufacturing rate (P), and facility setup cost (C), respectively.

Example 4.13

With a reduction in K, V value reduces. For better space utilization, the following options might be explored:

 i. Increase in y
 ii. Increase in P
 iii. Simultaneous increase in y and P

Now, for simultaneous increase in y and P:

$$y_1 = y\left(1 + \frac{Z}{100}\right)$$

$$P_1 = P\left(1 + \frac{S}{100}\right)$$

And, also:

$$K_1 = K\left(1 - \frac{W}{100}\right)$$

Now:

$$\sqrt{\frac{2CK_1D(1 - D / y_1P_1)}{I(I + K_1)}} = \sqrt{\frac{2CKD(1 - D / yP)}{I(I + K)}}$$

Or:

$$\frac{K_1(1 - D / y_1P_1)}{(I + K_1)} = \frac{K(1 - D / yP)}{(I + K)}$$

Or:

$$1 - \frac{D}{y_1P_1} = \frac{K(1 - D / yP)(I + K_1)}{K_1(I + K)}$$

Or:

$$\frac{D}{y_1P_1} = \frac{K_1(I + K) - K(1 - D / yP)(I + K_1)}{K_1(I + K)}$$

Or:

$$\frac{y_1P_1}{D} = \frac{K_1(I + K)}{K_1(I + K) - K(1 - D / yP)(I + K_1)}$$

Or:

$$y_1P_1 = \frac{DK_1(I + K)}{K_1(I + K) - K(1 - D / yP)(I + K_1)}$$

Or:

$$\left(1 + \frac{Z}{100}\right)\left(1 + \frac{S}{100}\right) = \frac{(D / yP)(I + K)K(1 - W / 100)}{K(I + K)(1 - W / 100) - K(1 - D / yP)\{I + K(1 - W / 100)\}}$$

Or:

$$\left(1 + \frac{Z}{100}\right)\left(1 + \frac{S}{100}\right) = \frac{(D / yP)(I + K)(1 - W / 100)}{(I + K)(1 - W / 100) - (1 - D / yP)\{I + K(1 - W / 100)\}} \tag{4.27}$$

Now each option may be explored.

TABLE 4.32
Values of Z(y) Corresponding to W (Backordering Cost)

S. No.	W	$Z = 100\left[\dfrac{(D/yP)(I+K)(1-W/100)}{(I+K)(1-W/100)-(1-D/yP)\{I+K(1-W/100)\}} - 1\right]$
1	4	0.82
2	8	1.72
3	12	2.72
4	16	3.85

a. Increase in y

$S = 0$, and from Eq. (4.27):

$$\left(1+\frac{Z}{100}\right) = \frac{(D/yP)(I+K)(1-W/100)}{(I+K)(1-W/100)-(1-D/yP)\{I+K(1-W/100)\}}$$

Or:

$$Z = 100\left[\frac{(D/yP)(I+K)(1-W/100)}{(I+K)(1-W/100)-(1-D/yP)\{I+K(1-W/100)\}} - 1\right]$$

For the given operational factors:

Proportion of acceptable components in a manufacturing batch, y = 0.95
Annual backordering cost per unit, K = ₹100
Manufacturing rate in units per year, P = 1250
Demand rate in units per year, D = 750
Annual production-inventory carrying cost per unit, I = ₹50
Facility setup cost, C = ₹75

Values of Z related to the proportion of acceptable components, are obtained in Table 4.32 corresponding to W related to the K.

b. Increase in P

$Z = 0$, and from Eq. (4.27):

$$\left(1+\frac{S}{100}\right) = \frac{(D/yP)(I+K)(1-W/100)}{(I+K)(1-W/100)-(1-D/yP)\{I+K(1-W/100)\}}$$

Or:

$$S = 100\left[\frac{(D/yP)(I+K)(1-W/100)}{(I+K)(1-W/100)-(1-D/yP)\{I+K(1-W/100)\}} - 1\right]$$

It can be observed that similar results would be available, i.e., concerning the previous option. However, the level of efforts needed to implement the specific option would be different and also varying from company to company.

TABLE 4.33

Combination of Z and S for W = 15

S. No.	Z	S
1	0.5	3.04
2	1.0	2.53
3	1.5	2.02
4	2.0	1.52
5	2.5	1.03

c. Simultaneous increase in y and P

For W = 15, various combinations of Z and S have been provided in Table 4.33 with the use of Eq. (4.27) considering the given data, such as:

$$\left(1+\frac{Z}{100}\right)\left(1+\frac{S}{100}\right)=1.0355329$$

Example 4.14

With an increase in the annual production-inventory carrying cost per unit (I), the maximum production-inventory level concerning acceptable components (V) reduces. For better space utilization, the following options may be explored:

a. Increase in y
b. Increase in P
c. Simultaneous increase in y and P

Now, for simultaneous increase in y and P:

$$y_1 = y\left(1+\frac{Z}{100}\right)$$

$$P_1 = P\left(1+\frac{S}{100}\right)$$

And, also:

$$I_1 = I\left(1+\frac{W}{100}\right)$$

Now:

$$\sqrt{\frac{2CKD(1-D/y_1P_1)}{I_1(I_1+K)}}=\sqrt{\frac{2CKD(1-D/yP)}{I(I+K)}}$$

Or:

$$\frac{(1-D/y_1P_1)}{I_1(I_1+K)}=\frac{(1-D/yP)}{I(I+K)}$$

Or:

$$1 - \frac{D}{y_1 P_1} = \frac{I_1(I_1 + K)(1 - D/yP)}{I(I + K)}$$

Or:

$$\frac{D}{y_1 P_1} = \frac{I(I + K) - I_1(I_1 + K)(1 - D/yP)}{I(I + K)}$$

Or:

$$\frac{y_1 P_1}{D} = \frac{I(I + K)}{I(I + K) - I_1(I_1 + K)(1 - D/yP)}$$

Or:

$$\left(1 + \frac{Z}{100}\right)\left(1 + \frac{S}{100}\right) = \frac{(D/yP)I(I + K)}{I(I + K) - I(1 + W/100)\{I(1 + W/100) + K\}(1 - D/yP)}$$

Or:

$$\left(1 + \frac{Z}{100}\right)\left(1 + \frac{S}{100}\right) = \frac{(D/yP)(I + K)}{(I + K) - (1 + W/100)(1 - D/yP)\{I(1 + W/100) + K\}} \quad (4.28)$$

Now each option may be explored.

a. Increase in y

$S = 0$, and from Eq. (4.28):

$$Z = 100\left[\frac{(D/yP)(I + K)}{(I + K) - (1 + W/100)(1 - D/yP)\{I(1 + W/100) + K\}} - 1\right]$$

Above equation can be used considering the data given in the previous example. The values of Z related to the proportion of acceptable components in a manufacturing lot, are provided in Table 4.34 corresponding to W values related to the annual inventory carrying cost per unit (I). The Z values increase with an upward variation in W, and also relatively more sensitive toward higher values of W. While

---

TABLE 4.34

Values of Z(y) Corresponding to W (Inventory Carrying Cost)

S. No.	W	$Z = 100\left[\dfrac{(D/yP)(I + K)}{(I + K) - (1 + W/100)(1 - D/yP)\{I(1 + W/100) + K\}} - 1\right]$
1	1	0.79
2	2	1.59
3	3	2.41
4	4	3.24
5	5	4.10

TABLE 4.35

Combination of Z and S for W = 10

S. No.	Z	S
1	1	7.59
2	2	6.53
3	3	5.50
4	4	4.48
5	5	3.49

implementing this option, the maximum y parameter cannot exceed 1, i.e., the increased y should be less than or equal to 1.

b. Increase in P

$Z = 0$, and from Eq. (4.28):

$$S = 100 \left[\frac{(D/yP)(I+K)}{(I+K) - (1+W/100)(1-D/yP)\{I(1+W/100)+K\}} - 1 \right]$$

It may be observed that similar results (i.e., related to the previous option) would be available. However, while implementing the present option, the required effort level can differ with reference to specific organization.

c. Simultaneous increase in y and P

From Eq. (4.8):

$$\left(1 + \frac{Z}{100}\right)\left(1 + \frac{S}{100}\right) = \frac{(D/yP)(I+K)}{(I+K) - (1+W/100)(1-D/yP)\{I(1+W/100)+K\}}$$

With the given data, and for $W = 10$:

$$\left(1 + \frac{Z}{100}\right)\left(1 + \frac{S}{100}\right) = 1.0866283$$

Now, different combinations of Z and S are provided in Table 4.35.

Along with the backorders and the subsequent rigorous entrepreneurial applications, quality inclusion has been described. This is followed by a variation in the proportion of acceptable components in a manufacturing batch and the related analysis. Interaction of different operational factors has also been analyzed with the help of appropriate options and suitable examples.

5 Conclusion

In this chapter, certain useful results have been provided to conclude. Also, the innovation efforts are discussed, and a future scope is envisioned.

5.1 USEFUL RESULTS

With reference to the various manufacturing parameters, certain conclusive comments are given.

5.1.1 IN THE CONTEXT OF SPACE CONSIDERATION

Some useful results are sequenced as follows:

1. Space consideration has been analyzed in sufficient detail in the context of manufacturing entrepreneurship. A wide variety of industrial products and components exists in terms of size and shape. Therefore, the space requirement differs from case to case. However, the generated maximum production-inventory level can be analyzed better in order to have a suitable generalization.
2. Maximum production-inventory level (V) in the manufacturing cycle can be expressed as:

$$\sqrt{\frac{2CD(1 - D/P)}{I}}$$

where:
P = Manufacturing rate in units per year
D = Demand rate in units per year
C = Facility setup cost
I = Annual production-inventory carrying cost per unit

3. The maximum production-inventory level (V) in the manufacturing cycle may increase because of:
   a. Demand rate reduction
   b. Facility setup cost increase
   c. Inventory carrying cost reduction
   d. Production rate increase
4. If the value of V increases, there is a requirement for additional space. This (i.e., the arrangement of space) might not be easier in many cases. However, if the V value can be brought to the original level by altering another feasible operational feature, a considerable success can be achieved in terms of manufacturing entrepreneurship.

5. With the additional parameter as follows:

y = Proportion of acceptable components in a manufacturing batch

Maximum production-inventory level concerning acceptable components (V) can be expressed as:

$$\sqrt{\frac{2CD(1-D/yP)}{I}}$$

6. With

W = % reduction in y

% reduction in V,

$$=\left[1-\sqrt{\frac{1-D/\{yP(1-W/100)\}}{(1-D/yP)}}\right]*100$$

7. With

W = % increase in y

% increase in V,

$$=\left[\sqrt{\frac{1-D/\{yP(1+W/100)\}}{(1-D/yP)}}-1\right]*100$$

8. With the additional parameter as follows:

K = Annual backordering cost per unit

A maximum production-inventory level (V) in the manufacturing cycle can be expressed as:

$$\sqrt{\frac{2CKD(1-D/P)}{I(I+K)}}$$

9. With

W = % increase in K

% increase in V,

$$=\left[\sqrt{\frac{(I+K)(1+W/100)}{I+K(1+W/100)}}-1\right]*100$$

10. With

W = % reduction in K

% reduction in V,

$$=\left[1-\sqrt{\frac{(I+K)(1-W/100)}{I+K(1-W/100)}}\right]*100$$

11. Because of an increase in K, the V value also increases. To bring back the V value to an original level, the suitable options may include:
    a. Reduction in C
    b. Reduction in P
    c. Simultaneous reduction in C and P
12. With increase in the manufacturing rate, the V value increases. For the goal of similar V value, the following suitable remedial measures may be available:
    a. Reduction in C
    b. Reduction in K
    c. Simultaneous reduction in C and K
13. With a reduction in the holding cost, V value increases. The suitable options may include:
    a. Reduction in K
    b. Reduction in P
    c. Reduction in C
    d. Simultaneous reduction of K and P
    e. Simultaneous reduction of P and C
    f. Simultaneous reduction of C and K
    g. Simultaneous reduction of all the three factors, i.e., K, P, and C
14. With an increase in the facility setup cost, the V value increases. To deal with this, the following remedial measures may be considered:
    a. Reduction in K
    b. Reduction in P
    c. Simultaneous reduction in K and P
15. With

y = Proportion of acceptable components in a manufacturing batch

A maximum production-inventory level (V) including backorders can be expressed as:

$$\sqrt{\frac{2CKD(1 - D/yP)}{I(I + K)}}$$

5.2 INNOVATION EFFORTS

A company might deal with innovative products from the start. However, in most of the manufacturing organizations, innovative efforts run in parallel with the production of regular/routine items. Such efforts might be associated with:

a. Research and development related to new ideas
b. Design improvement
c. Change in production tooling
d. Change in process design

There is a possibility to include some of such efforts in terms of facility setup cost for a certain period.

5.2.1 MANUFACTURE OF INNOVATIVE ITEM ALONG WITH REGULAR PRODUCT

In the case where the innovative item is handled independently, initial additional investment also needs to be considered. In view of the time value of money, limited number of years/cycles may be planned for such an item initially. This can also depend on the life cycle of the product. However, when such an item is to be produced along with the routine/regular product, a maximum inventory level in the context of space requirement can play a certain role in the decision-making process.

With reference to Figure 2.4, the cycle time (T) is obtained as follows:

$$\frac{V}{D(1-D/P)}$$

To have a suitable approach, let subscript "1" refer to the regular product, and subscript "2" refer to the innovative product. Thus:

$$V_1 = TD_1(1-D_1/P_1) \tag{5.1}$$

$$V_2 = TD_2(1-D_2/P_2) \tag{5.2}$$

From Eq. (2.3), the total cost can be transformed as follows:

$$E = \frac{1}{2}[V_1I_1 + V_2I_2] + \left[\frac{C_1D_1(1-D_1/P_1)}{V_1} + \frac{C_2D_2(1-D_2/P_2)}{V_2}\right]$$

Substituting Eqs. (5.1) and (5.2):

$$E = \frac{T}{2}[D_1I_1(1-D_1/P_1) + D_2I_2(1-D_2/P_2)] + \frac{(C_1+C_2)}{T}$$

$$\frac{dE}{dT} = 0$$

indicates:

$$\frac{1}{2}[D_1I_1(1-D_1/P_1) + D_2I_2(1-D_2/P_2)] = \frac{(C_1+C_2)}{T^2}$$

Or

$$T^2 = \frac{2(C_1+C_2)}{[D_1I_1(1-D_1/P_1) + D_2I_2(1-D_2/P_2)]}$$

Thus, the optimal cycle time, T, is expressed as follows:

$$T = \sqrt{\frac{2(C_1+C_2)}{[D_1I_1(1-D_1/P_1) + D_2I_2(1-D_2/P_2)]}} \tag{5.3}$$

Example 5.1

For the regular item, the details are as follows:

Demand rate in units per year, $D_1 = 750$
Manufacturing rate in units per year, $P_1 = 1250$
Annual inventory carrying cost per unit, $I_1 = ₹50$
Facility setup cost, $C_1 = ₹75$

For the innovative item to be produced in the same manufacturing cycle along with the mentioned regular item, the details are given as follows:

Demand rate in units per year, $D_2 = 100$
Manufacturing rate in units per year, $P_2 = 1000$
Annual inventory carrying cost per unit, $I_2 = ₹60$
Facility setup cost, $C_2 = ₹90$

Now, from Eq. (5.3):

$$T = \sqrt{\frac{2(75 + 90)}{[15000 + 5400]}}$$

$$= \sqrt{\frac{330}{20400}}$$

$$= 0.127 \text{ year}$$

From Eq. (5.1):

$$V_1 = TD_1(1 - D_1 / P_1)$$

$$= 38.1$$

From Eq. (5.2):

$$V_2 = TD_2(1 - D_2 / P_2)$$

$$= 11.43$$

However, in terms of input operational factors also, the value for maximum inventory level for both the products can be obtained by substituting Eq. (5.3) in Eqs. (5.1) and (5.2), that is:

$$V_1 = D_1(1 - D_1 / P_1)\sqrt{\frac{2(C_1 + C_2)}{[D_1 I_1(1 - D_1 / P_1) + D_2 I_2(1 - D_2 / P_2)]}} \tag{5.4}$$

$$V_2 = D_2(1 - D_2 / P_2)\sqrt{\frac{2(C_1 + C_2)}{[D_1 I_1(1 - D_1 / P_1) + D_2 I_2(1 - D_2 / P_2)]}} \tag{5.5}$$

For the space requirement planning, it is necessary to consider the interaction of both the products, i.e., the regular and innovative product. After the production of an item ends, manufacturing of another item begins. Inventory build-up of second item is also associated with the consumption of the first item. Such aspects help in the estimation for the required space for both the items as well as appropriate sequencing of items in the context of manufacture.

While initiating an innovative product, quality issues may occur, and therefore the quality aspects may also be included in this approach.

With reference to Figure 3.2, the cycle time (T) is obtained as follows:

$$\frac{V}{D(1-D/yP)}$$

Now, for regular and innovative products:

$$V_1 = TD_1(1-D_1/y_1P_1) \tag{5.6}$$

$$V_2 = TD_2(1-D_2/y_2P_2) \tag{5.7}$$

where:

y_1 = Proportion of acceptable components in a manufacturing batch for regular item

y_2 = Proportion of acceptable components in a manufacturing batch for innovative item

From Eq. (3.3), the total cost can be transformed as follows:

$$E = \frac{1}{2}[V_1I_1 + V_2I_2] + \left[\frac{C_1D_1(1-D_1/y_1P_1)}{V_1} + \frac{C_2D_2(1-D_2/y_2P_2)}{V_2}\right]$$

Substituting Eqs. (5.6) and (5.7):

$$E = \frac{T}{2}[D_1I_1(1-D_1/y_1P_1) + D_2I_2(1-D_2/y_2P_2)] + \frac{(C_1+C_2)}{T}$$

$$\frac{dE}{dT} = 0$$

indicates:

$$\frac{1}{2}[D_1I_1(1-D_1/y_1P_1) + D_2I_2(1-D_2/y_2P_2)] = \frac{(C_1+C_2)}{T^2}$$

Or:

$$T^2 = \frac{2(C_1+C_2)}{[D_1I_1(1-D_1/y_1P_1) + D_2I_2(1-D_2/y_2P_2)]}$$

Thus the optimal cycle time is expressed as follows:

$$T = \sqrt{\frac{2(C_1+C_2)}{[D_1I_1(1-D_1/y_1P_1) + D_2I_2(1-D_2/y_2P_2)]}} \tag{5.8}$$

Example 5.2

Consider the data of Example 5.1. Additionally, for the regular item:

$$y_1 = 0.99$$

And, for the innovative item:

$$y_2 = 0.95$$

Now, from Eq. (5.8):

$$T = 0.128 \text{ year}$$

From Eq. (5.6):

$$V_1 = TD_1(1 - D_1 / y_1 P_1)$$

$$= 37.82$$

From Eq. (5.7):

$$V_2 = TD_2(1 - D_2 / y_2 P_2)$$

$$= 11.45$$

With the inclusion of quality aspects, the manufacturing cycle time has relatively increased. Depending on the input operational factors, the value of V_1 comparatively decreases, whereas the value of V_2 comparatively increases.

In terms of input operational factors also, the value for maximum inventory level for both the products can be obtained by substituting Eq. (5.8) in Eqs. (5.6) and (5.7), that is:

$$V_1 = D_1(1 - D_1 / y_1 P_1) \sqrt{\frac{2(C_1 + C_2)}{[D_1 I_1(1 - D_1 / y_1 P_1) + D_2 I_2(1 - D_2 / y_2 P_2)]}} \tag{5.9}$$

$$V_2 = D_2(1 - D_2 / y_2 P_2) \sqrt{\frac{2(C_1 + C_2)}{[D_1 I_1(1 - D_1 / y_1 P_1) + D_2 I_2(1 - D_2 / y_2 P_2)]}} \tag{5.10}$$

When the focus is on space aspects only, above equations can directly be used considering input operational factors, i.e., without obtaining a manufacturing cycle time.

When backorders are also to be incorporated, the total related cost is given by Eq. (4.7) as:

$$E = \frac{C}{T} + \frac{V^2(I + K)}{2TD(1 - D / P)} + \frac{KTD(1 - D / P)}{2} - VK$$

With an objective of minimizing the total related cost, differentiating partially with respect to V and equating to zero,

$$\frac{V(I + K)}{TD(1 - D / P)} = K$$

Or:

$$V = TD(1 - D / P)K / (I + K)$$

Now:

$$V_1 = TD_1(1 - D_1 / P_1)K_1 / (I_1 + K_1) \tag{5.11}$$

$$V_2 = TD_2(1 - D_2 / P_2)K_2 / (I_2 + K_2) \tag{5.12}$$

Total related cost given by Eq. (4.8) can be transformed as follows:

$$E = \frac{C_1 K_1 D_1 (1 - D_1 / P_1)}{V_1 (I_1 + K_1)} + \frac{C_2 K_2 D_2 (1 - D_2 / P_2)}{V_2 (I_2 + K_2)} + \frac{V_1 I_1}{2} + \frac{V_2 I_2}{2}$$

Substituting Eqs. (5.11) and (5.12)

$$E = \frac{(C_1 + C_2)}{T} + \frac{T}{2}\left[\frac{I_1 D_1 (1 - D_1 / P_1)K_1}{(I_1 + K_1)} + \frac{I_2 D_2 (1 - D_2 / P_2)K_2}{(I_2 + K_2)}\right]$$

$$\frac{dE}{dT} = 0$$

indicates:

$$T^2 = \frac{2(C_1 + C_2)}{[\{I_1 D_1 (1 - D_1 / P_1)K_1 / (I_1 + K_1)\} + \{I_2 D_2 (1 - D_2 / P_2)K_2 / (I_2 + K_2)\}]}$$

Thus, the optimal cycle time is expressed as follows:

$$T = \sqrt{\frac{2(C_1 + C_2)}{[\{I_1 D_1 (1 - D_1 / P_1)K_1 / (I_1 + K_1)\} + \{I_2 D_2 (1 - D_2 / P_2)K_2 / (I_2 + K_2)\}]}} \tag{5.13}$$

Example 5.3

For the regular item, the details are given as follows:

    Demand rate in units per year, $D_1 = 750$
    Manufacturing rate in units per year, $P_1 = 1250$
    Annual inventory carrying cost per unit, $I_1 = ₹50$
    Facility setup cost, $C_1 = ₹75$
    Annual backordering cost per unit, $K_1 = ₹100$

For the innovative item to be produced in the same manufacturing cycle along with the mentioned regular item, the details are given as follows:

    Demand rate in units per year, $D_2 = 100$
    Manufacturing rate in units per year, $P_2 = 1000$

Annual inventory carrying cost per unit, $I_2 = ₹60$
Annual backordering cost per unit, $K_2 = ₹120$
Facility setup cost, $C_2 = ₹90$

Now, from Eq. (5.13):

$$T = \sqrt{\frac{2(75+90)}{[10000+3600]}}$$

$$= \sqrt{\frac{330}{13600}}$$

$$= 0.1558 \text{ year}$$

From Eq. (5.11):

$$V_1 = TD_1(1 - D_1 / P_1)K_1 / (I_1 + K_1)$$

$$= 31.15$$

From Eq. (5.12):

$$V_2 = TD_2(1 - D_2 / P_2)K_2 / (I_2 + K_2)$$

$$= 9.346$$

Since backorders are allowed, the values for V_1 and V_2 reduce.
In terms of input operational factors, Eq. (5.13) can also be substituted in Eqs. (5.11) and (5.12) in order to obtain V_1 and V_2 respectively, such as:

$$V_1 = \frac{D_1(1 - D_1 / P_1)K_1}{(I_1 + K_1)} \sqrt{\frac{2(C_1 + C_2)}{[\{I_1D_1(1 - D_1 / P_1)K_1 / (I_1 + K_1)\} + \{I_2D_2(1 - D_2 / P_2)K_2 / (I_2 + K_2)\}]}}$$

(5.14)

And,

$$V_2 = \frac{D_2(1 - D_2 / P_2)K_2}{(I_2 + K_2)} \sqrt{\frac{2(C_1 + C_2)}{[\{I_1D_1(1 - D_1 / P_1)K_1 / (I_1 + K_1)\} + \{I_2D_2(1 - D_2 / P_2)K_2 / (I_2 + K_2)\}]}}$$

(5.15)

When quality aspects are also to be incorporated along with the backorders, the total related cost is given by Eq. (4.22) such as:

$$E = \frac{C}{T} + \frac{V^2(I + K)}{2TD(1 - D / yP)} + \frac{KTD(1 - D / yP)}{2} - VK$$

With an objective of minimizing the total related cost, differentiating partially with respect to V and equating to zero,

$$\frac{V(I + K)}{TD(1 - D / yP)} = K$$

Or:

$$V = TD(1 - D / yP)K / (I + K)$$

Now:

$$V_1 = TD_1(1 - D_1 / y_1P_1)K_1 / (I_1 + K_1) \tag{5.16}$$

$$V_2 = TD_2(1 - D_2 / y_2P_2)K_2 / (I_2 + K_2) \tag{5.17}$$

where
  y_1 = Proportion of acceptable components in a manufacturing batch for regular item
  y_2 = Proportion of acceptable components in a manufacturing batch for innovative item

Total related cost given by Eq. (4.23) can be transformed as follows:

$$E = \frac{C_1K_1D_1(1 - D_1 / y_1P_1)}{V_1(I_1 + K_1)} + \frac{C_2K_2D_2(1 - D_2 / y_2P_2)}{V_2(I_2 + K_2)} + \frac{V_1I_1}{2} + \frac{V_2I_2}{2}$$

Substituting Eqs. (5.16) and (5.17)

$$E = \frac{(C_1 + C_2)}{T} + \frac{T}{2}\left[\frac{I_1D_1(1 - D_1 / y_1P_1)K_1}{(I_1 + K_1)} + \frac{I_2D_2(1 - D_2 / y_2P_2)K_2}{(I_2 + K_2)}\right]$$

$$\frac{dE}{dT} = 0$$

indicates:

$$T^2 = \frac{2(C_1 + C_2)}{[\{I_1D_1(1 - D_1 / y_1P_1)K_1 / (I_1 + K_1)\} + \{I_2D_2(1 - D_2 / y_2P_2)K_2 / (I_2 + K_2)\}]}$$

Thus the optimal cycle time is expressed as follows:

$$T = \sqrt{\frac{2(C_1 + C_2)}{[\{I_1D_1(1 - D_1 / y_1P_1)K_1 / (I_1 + K_1)\} + \{I_2D_2(1 - D_2 / y_2P_2)K_2 / (I_2 + K_2)\}]}} \tag{5.18}$$

After getting the cycle time from the above expression, the values for V_1 and V_2 can be obtained from Eqs. (5.16) and (5.17), respectively.

Example 5.4

Consider the data of Example 5.3. Additionally, for the regular item:

$$y_1 = 0.99$$

And, for the innovative item:

$$y_2 = 0.95$$

Now, from Eq. (5.18):

$$T = 0.1568 \text{ year}$$

With the inclusion of quality aspects, the manufacturing cycle time has relatively increased.

Now, from Eq. (5.16):

$$V_1 = TD_1(1 - D_1 / y_1P_1)K_1 / (I_1 + K_1)$$

$$= 30.88$$

And, from Eq. (5.17):

$$V_2 = TD_2(1 - D_2 / y_2P_2)K_2 / (I_2 + K_2)$$

$$= 9.351$$

Depending on the input operational factors, the value of V_1 comparatively decreases, whereas the value of V_2 comparatively increases. Such information helps in knowing the changed occupied space estimate.

While making innovation efforts along with the regular manufacturing output, the effects of the changed levels of maximum production-inventory might be examined in an organization considering the specific operational factors. This helps eventually in the space requirement planning for respective items, as the shape and size may differ for the regular and innovative products.

In the above description, an innovative product is handled jointly along with the regular item. However, it is also possible to analyze an innovative item independently depending on the nature of business.

5.2.2 HANDLING INNOVATIVE ITEM INDEPENDENTLY

Certain type of entrepreneurs might be interested in an innovative item from the start, and therefore an independent analysis may be useful in this specific context. This may also be beneficial in linking the inventory-related costs among other factors with the product innovation. In addition to the innovation efforts and related expenditure, other factors may include:

a. Predicted customer demands
b. Production capacity
c. Fixed setup cost

Depending on the short product life cycle, the formulation and analysis can include a limited planned horizon such as the 2 or 3 years or certain number of months.

Let:

D = Total demand during the planned number of months (T)
P = Total production capacity (rate) for the planned number of months (T)
T = Planned horizon in months

H = Carrying cost in ₹/unit-month
t = Planned cycle time for frequent production activity in months
A = Fixed setup cost
I = Product innovation cost apportioned for one cycle concerned with the time
  value of money
M = Maximum inventory concerning one cycle related a limited planning
  horizon

To start the discussion, Figure 5.1 shows the frequent production activity during the planned horizon.

Also, Figure 5.2 represents one cycle along with the parameters.

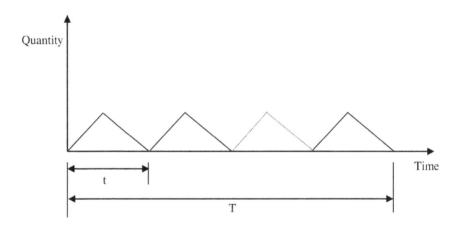

FIGURE 5.1   Frequent production activity during the planned horizon.

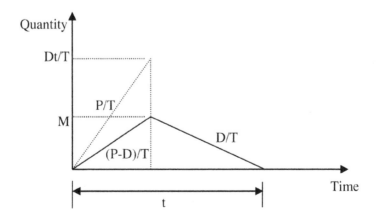

FIGURE 5.2   Depiction of one cycle.

Now:
 Production quantity per cycle,

$$= \frac{Dt}{T}$$

And:

$$\frac{M}{(P-D)/T} = \frac{Dt/T}{P/T}$$

Or:

$$M = \frac{(P-D)}{T} \cdot \frac{Dt}{P}$$

Or:

$$M = \frac{Dt(1-D/P)}{T} \qquad (5.19)$$

Inventory cost during the cycle,

$$= \frac{M}{2} \cdot H \cdot t$$

And the total inventory cost for the planning horizon T,

$$= \frac{M}{2} \cdot Ht \cdot \frac{T}{t}$$

$$= \frac{M}{2} \cdot HT$$

Substituting Eq. (5.19):
 Total inventory cost for the planning horizon T,

$$= \frac{Dt(1-D/P)}{2T} \cdot HT$$

$$= \frac{DHt(1-D/P)}{2} \qquad (5.20)$$

Total setup cost,

$$= \frac{T}{t} \cdot A \qquad (5.21)$$

Total innovation cost,

$$= \frac{T}{t} \cdot I \qquad (5.22)$$

Adding Eqs (5.20), (5.21), and (5.22), total relevant cost, E, is given as follows:

$$E = \frac{T(A+I)}{t} + \frac{DHt(1-D/P)}{2}$$

Now, to obtain the optimal value of t:

$$\frac{dE}{dt} = \frac{-T(A+I)}{t^2} + \frac{DH(1-D/P)}{2} = 0$$

Or:

$$\frac{DH(1-D/P)}{2} = \frac{T(A+I)}{t^2}$$

Or:

$$t = \sqrt{\frac{2T(A+I)}{DH(1-D/P)}} \qquad (5.23)$$

Example 5.5

In order to illustrate, consider:

Planned horizon in months, T = 24 months
Total demand during the planned number of months (T), D = 12,000
Total production capacity (rate) for the planned number of months (T),
   P = 14,400
Carrying cost, H = ₹2 per unit-month
Fixed setup cost, A = ₹300
Product innovation cost apportioned for one cycle, I = ₹1000

From Eq. (5.23), optimal cycle time in months can be evaluated as:

$$t = \sqrt{\frac{2 \times 24 \times 1300}{12000 \times 2[1-(12000/14400)]}}$$

Or:

$$t = 3.9497 \text{ months}$$

$$\approx 4 \text{ months}$$

In the context of space consideration, the value of M can be obtained by substituting Eq. (5.23) in Eq. (5.19):

$$M = \frac{D(1 - D/P)}{T} \sqrt{\frac{2T(A + I)}{DH(1 - D/P)}}$$

Or:

$$M = \sqrt{\frac{2(A + I)D(1 - D/P)}{TH}} \tag{5.24}$$

Considering the given data:

$$M = \sqrt{\frac{2 \times 1300 \times 12000[1 - (12000/14400)]}{24 \times 2}}$$

$$= 329.14$$

Suitable space needs to be arranged for implementation of such scheme of manufacture. After knowing the maximum inventory level, an idea can be obtained concerning the space requirement that helps in suitable planning for generation of space. However, the space requirement will depend on:

i. Type of industry
ii. Product type
iii. Design of the product
iv. Nature of business
v. Stacking method for the concerned product
vi. Need for movement of handling equipment
vii. Need for movement of human resources
viii. Requirement for quality inspection on the shop floor
ix. Requirement for keeping the inspection equipment temporarily

5.3 FUTURE SCOPE

In the present work, single product is largely considered in the context of manufacturing entrepreneurship. However, with reference to the innovative efforts, the two items, i.e., regular and innovative item, are handled in a manufacturing cycle. To provide better generalization approaches, the future scope lies in formulating multiple items produced in a cycle and the implication for space requirement planning in a production/manufacturing organization.

In addition, the significant scope also lies in highlighting various applications of the concepts suggested in this book. For instance, a production line, as shown in Figure 5.3, might be perfectly balanced, and also because of the nature of product, the final item is available at the end of the line in a relatively short time.

Now the manufacturing or production rate can be envisioned as the overall rate of production in terms of availability at the end of the line. How quickly or at what rate these items are lifted from that place of final availability may be imagined as the relevant consumption or demand rate in the present context of space implication in any manufacturing organization.

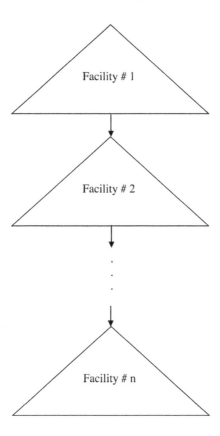

FIGURE 5.3   Production line with a group of facilities.

Depending on the organization and the nature of product, such line may be associated with:

  i. Production
 ii. Assembly
iii. Packaging
 iv. Filling

From such line, the items are taken to other places varying with specific case as shown in Figure 5.4.

The places where the items are brought from the end of production line, such as immediate/adjacent destinations, might be as follows:

  a. Testing
  b. Packaging
  c. Warehouse

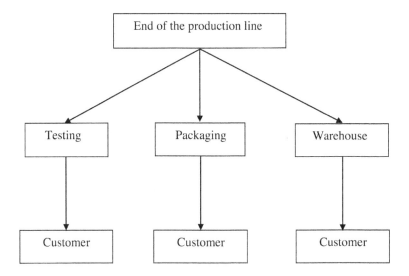

FIGURE 5.4 Various destinations with respect to specific case.

In case of the testing, it may be an adjacent as well as immediate/next destination. It may be in the same premises also. Packaging might be in the same premises as well as elsewhere depending on the need and economic consideration among others. Warehouse may also be near to or far away from the production facility. The warehouse might be the central warehouse or may be replaced by distribution center. It may also happen that the items are dispatched from the central warehouse to the distribution centers and finally to the customers. Eventually, the end destination is the customer.

For an analysis, specific case might be developed. Future scope includes the development of a wide variety of practical cases and subsequent analyses. As represented by Figure 5.5, different cases might be evolved. Such cases may include:

a. For some of the products, packaging might not be essential or significant. And therefore, the items may be taken to the customer organizations directly.
b. In some cases, packaging is necessary after testing or inspection, and then the items are taken to the customer organizations.
c. After the testing and packaging, it may be necessary to dispatch the items to certain warehouse and then finally to the customer companies.

However, different situations are not limited to only those mentioned here. From the central warehouse, the products may be transported to a regional warehouse or distribution center, and also to dealers/retailers. From the dealers/retailers, the items can be purchased by individual consumers finally.

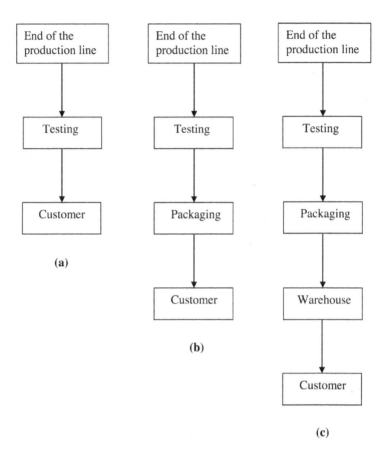

FIGURE 5.5   Different cases.

In the context of manufacturing entrepreneurship, various interfaces need to be visualized along with a perception of the related operational factors. For example, the packaging might be performed at the manufacturing plant or in the warehouse also, depending on the nature of business. In either case, an arrival rate at the packaging station/location can be visualized as the replenishment rate (that might be analogous to the production or manufacturing rate for the purpose of analysis). At an appropriate interface, the rate at which the packaging operation is performed might be analogous to the consumption or demand rate for the purpose of analysis.

Utility of the suggested concepts can be emphasized at an organizational level as well as at the facility level in the context of space requirement planning among other issues. For instance, the innovation efforts have been mentioned before, along with the clubbing of innovative item with the regular item. For an assessment of space requirement, consider the production-inventory build-up, as shown in Figure 5.6.

When the manufacture of regular item ends, it can begin for the innovative item. For the space requirement, status of both the items should be carefully considered with respect to time.

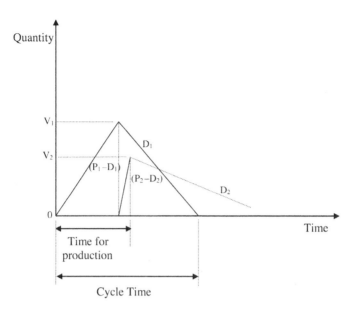

FIGURE 5.6   Production-inventory build-up of two items.

During the time interval as follows:

After the end of production of the first item and before the end of production
of the second item

Inventory starts depleting for the first item, and the inventory build-up is happening
for the second item. At any point of time, inventory status of both the items needs to
be considered. For the assessment of total required space, the shape and size of both
the items should be included if these differ considerably.

For the next time interval, that is:

After the end of the total time for production and before the end of cycle time

Inventory for both the items are depleting, and accordingly, the status should be
considered in addition to other aspects.

A detailed analysis may be made concerning a specific case, particularly for
knowing the total maximum inventory level related to both the items. If there is
a need in the context of total space requirement, sequencing of items may also be
taken care for production purpose, in case where it is beneficial.

As discussed before, the suggested concepts in this book are useful at facility/shop
floor level as well as at organizational level. A vision for future scope also includes
a better review of available space at different interfaces such as:

i. Predecessor and successor facility
ii. Production line and assembly line

iii. Manufacture and inspection/testing
iv. Manufacture and packaging
 v. Packaging and loading area for transport
vi. Unloading area and warehouse

Perception of operational factors at different interfaces helps a lot in the review and analysis concerning the availability and requirement of space. Depending on the business and frequency for material movement as well as temporary storage, a shift in some of the identified facilities can also be thought. An investment needed in reinstalling few identified facilities in the layout can be compared with the potential benefits in the short term as well as long term.

Advanced Reading I
A Novel Approach to Look at 'Green' Facility Setup Cost

I.1   INTRODUCTION

Traditionally products are manufactured in batches. However, just-in-time (JIT) concept is also applied depending on the context. JIT focuses on smaller batch size and lower setup costs. With the environmental consciousness, internal as well as external movement of unnecessarily smaller batch size might not be preferred. This is because of energy consumption and pollution among other aspects which are not environment friendly. With the manufacturing rate increase, a suitable lot size can be produced in comparatively less amount of time. The material movement might happen timely along with its being energy/environment friendly. Therefore, significant scope still exists for appropriate batch size and traditional balancing of the costs such as:

a. Facility setup cost
b. Inventory holding cost

Other additional parameters which are taken into consideration are:

i. Demand rate
ii. Production rate
iii. Production time cost

Flexibility of demand and production rate has been discussed (Sharma 2008). Production rate is also decreased for few purposes such as dealing with shelf life (Sarker and Babu 1993; Sharma 2004, 2006, 2007, 2009; Silver 1990, 1995) among other reasons. Often the companies are in the mode of increasing their production rates and therefore such an increase appears to be an interesting topic for analysis. Furthermore, as it is mentioned before, appropriate batch size can be in a better synergy from environment/energy point of view. However, it is also important to know whether it is economical too. This chapter is devoted to creating an upper bound for facility setup cost so that the company can know under what condition, an overall cost reduction can also be ensured in addition to the discussed benefits of an increase in production rate. While doing so, if environmentally positive scenario can also be observed by the organization and associated stakeholders in addition to economic benefits, the proposed approach can be termed as a step toward 'green' facility setup cost.

In a truly flexible production-inventory system, it is possible to vary the manufacturing rate in following ways:

   i. Upward variation
  ii. Downward variation

The analysis is also extended for downward variation in the manufacturing rate in case where a lower bound for facility setup cost is convenient for implementation. The chapter also discusses illustrative examples along with a step-wise implementation procedure briefly. The cases including backorders are also discussed appropriately.

I.2   MANUFACTURING RATE VARIATION

Assume:

A = Setup cost ($)
C = Production time cost ($per year)
D = Annual demand
H = Annual holding cost per unit ($per unit)
P = Production rate (units per year)

In a batch manufacturing environment, the annual costs are as follows:
   Setup cost,

$$= DA / Q \tag{1}$$

Production cost,

$$= CD / P \tag{2}$$

Inventory holding cost,

$$= (P - D)HQ / 2P \tag{3}$$

Adding expressions (1), (2), and (3), total relevant cost:

$$E = (DA / Q) + (CD / P) + (1 - D / P)HQ / 2 \tag{4}$$

It can be shown that the optimal relevant cost:

$$E^* = (CD / P) + \sqrt{2DA(1 - D / P)H} \tag{5}$$

I.2.1   UPWARD VARIATION

As the upward variation in the production rate (Figure I.1) and its effects are to be analyzed, let:

P_1 = Increased production rate (units per year)

Now the relevant cost is given as:

$$E_1^* = (CD / P_1) + \sqrt{2DA(1 - D / P_1)H} \tag{6}$$

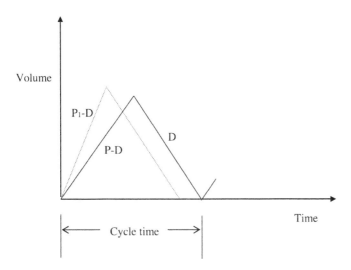

FIGURE I.1   Manufacturing rate increase.

Subtracting Eq. (6) from (5), and also $(E^* - E_1^*) > 0$ shows:

$$A < \frac{DC^2}{2H} \left[\frac{(1/P) - (1/P_1)}{\sqrt{(1 - D/P_1)} - \sqrt{(1 - D/P)}} \right]^2 \qquad (7)$$

The facility setup cost has an important role in deciding whether production rate enhancement is useful from cost point of view in addition to the environmental positive aspects.

In order to consider a suitable example, assume:

A = $900
C = $8000
D = 360 units
H = $90
P = 480 units
P_1 = 500 units

From the expression (7), the assumed facility setup cost satisfies the condition and therefore the green benefits of production rate increase can be achieved with cost reduction too.

From Eq. (5), $E^* = \$9818.38$.

And from Eq. (6), $E_1^* = \$9800.99$.

Since the focus is on facility setup cost and companies often make sincere efforts to reduce this cost, it is reasonable to observe the effects on the total cost with reduction in setup cost (Table I.1).

The reduction in the cost is more sensitive as the setup cost is decreased further.

TABLE I.1

Variation in the Total Cost with Setup Cost Reduction (Increased Manufacturing Rate)

A ($)	E_1*($)
900	9800.99
890	9778.48
880	9755.84
870	9733.07
860	9710.17
850	9687.14

I.2.2 DOWNWARD VARIATION

Refering to Eq. (4), we can see that the optimal manufacturing batch size is:

$$Q^* = \sqrt{2DA / [(1 - D / P)H]} \qquad (8)$$

In the previous case, i.e., the upward manufacturing rate variation, batch size became comparatively smaller. This may be appropriate for some companies. However, in practice, suitability of batch size might depend on several factors such as material handling equipment and existing vessel size among other features. Therefore, the discussion can be extended for manufacturing rate decrease (Figure I.2) also, in case of which, the batch size becomes comparatively larger.

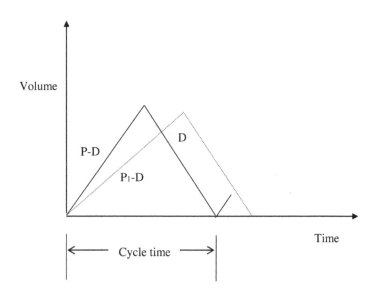

FIGURE I.2   Manufacturing rate decrease.

TABLE I.2

Variation in the Total Cost with Setup Cost Reduction (Decreased Manufacturing Rate)

A ($)	$E_1^*($)$
1000	10014.13
990	9995.32
980	9976.41
970	9957.40
960	9938.30
950	9919.09

Following the relevant steps of Section 2.1, the expression concerning a facility setup cost is:

$$A > \frac{DC^2}{2H} \left[\frac{(1/P_1) - (1/P)}{\sqrt{(1 - D/P)} - \sqrt{(1 - D/P_1)}} \right]^2 \tag{9}$$

Assume the basic data of previous example except the setup cost.

Now, the decreased rate of manufacture, $P_1 = 460$ units (say).

From the expression (9): $A > 922.12$. Therefore, consider $A = \$1000$.

From Eq. (5), $E^* = \$10,024.92$.

From Eq. (6), $E_1^* = \$10,014.13$.

Effects on the total cost with setup cost reduction are shown in Table I.2. In this case also, decrease in total cost is more sensitive as the setup cost is decreased.

As far as the practical implementation of the proposed approach is concerned, every company has to assess its operational parameters and other features, i.e., both quantitative and qualitative, in order to take suitable decision or determine appropriate policy (Sharma 2010).

The implementation procedure may be briefly mentioned as follows:

a. Estimate the facility setup cost.

b. If setup cost satisfies the expression (7), then enhance the manufacturing rate. Else go to (e).

c. Check whether an increase in the manufacturing rate is also beneficial from 'green' point of view, in addition to economic benefit. If it is true, inform all the stakeholders so that an implementation becomes smooth.

d. Effort should now be made for further setup cost reduction because of certain incentive.

e. If setup cost satisfies the expression (9), then decrease the rate of manufacture.

f. Check whether this reduction in the production rate can be visualized as 'green' also for ease in implementation.

g. Setup cost can now be decreased further because of the demonstrated incentive for the company.

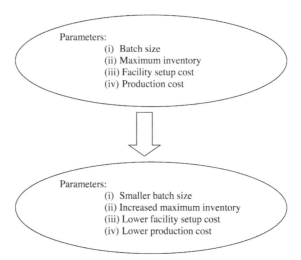

FIGURE I.3   Transformation regarding the increase in production rate.

Figures I.3 and I.4 show the transformation regarding allowable flexibility in production rate. Significant parameters are represented on the up side. Relative change is shown on the bottom side, i.e., the potential destination on implementation of a manufacturing rate variation after optimizing the then scenario. Lower/higher setup costs are indicative of upper/lower bounds respectively for economic transformation. With the proposed approach and its variants, it is possible to quantify an increase in maximum inventory (where applicable). Practical/managerial implication lies in checking for enough storage space for work-in-process/finished products and also such requirements for raw materials/input items. Production cost depends on the

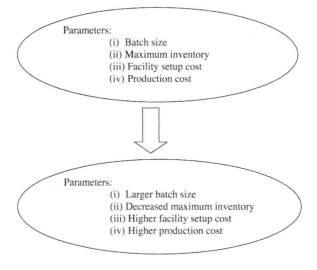

FIGURE I.4   Transformation regarding the decrease in production rate.

manufacturing time involved. Manufacturing time is lower/higher with an increased/decreased rate of manufacture respectively. Either smaller or larger batch size might prove to be appropriate batch size depending on the overall costs, environmental/energy benefits, and other qualitative/quantitative aspects pertaining to individual organizational culture, operations, and corporate social responsibility.

I.3   BACKORDERS

In the manufacturing inventory analysis, backorders also might be allowed. Figure I.5 shows a manufacturing rate increase when certain shortage quantities are backordered. Volume variation is depicted when rate of manufacture is increased from P to P_1.

In the context of backorders, Figure I.6 represents volume variation when a decrease in manufacturing rate is observed.

I.3.1   CASE WHEN THE SHORTAGES ARE FULLY BACKORDERED

In a separate analysis, costs have been mentioned (Sharma 2007). For the current scenario also, the total cost can be transformed and expressed as follows:

$$E^* = (CD / P) + \sqrt{2DA(1 - D / P)HK / (K + H)} \tag{10}$$

where K = Annual shortage cost per unit.

For the upward production rate variation (following similar procedure), it can be shown that:

$$A < \frac{DC^2(K + H)}{2HK} \left[\frac{(1 / P) - (1 / P_1)}{\sqrt{(1 - D / P_1)} - \sqrt{(1 - D / P)}} \right]^2 \tag{11}$$

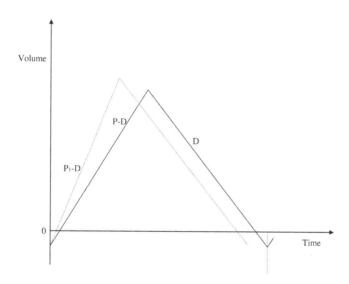

FIGURE I.5   Manufacturing rate increase (with backorders).

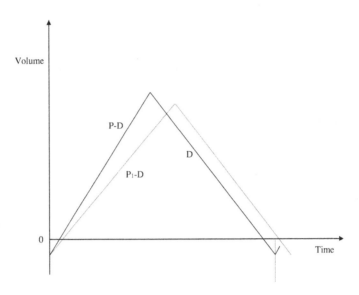

FIGURE I.6   Manufacturing rate decrease (with backorders).

Consider the following data:

A = $900
C = $8000 per year
D = 360 units
H = $90
P = 480 units/year
P_1 = 500 units/year
K = $50

The costs are calculated as follows:

E* = $8281.92
E_1* = $8174.95

Table I.3 shows effect on the cost with a reduction in setup cost.

Cost reduction is more sensitive with reduction in setup cost.

In case of the downward production rate variation, a necessary condition for economic scenario is as follows:

$$A > \frac{DC^2(K+H)}{2HK}\left[\frac{(1/P_1)-(1/P)}{\sqrt{(1-D/P)}-\sqrt{(1-D/P_1)}}\right]^2 \tag{12}$$

Considering the previous example, expression (12) indicates:

A > $2581.94

TABLE I.3
Revised Cost When Shortages Are Fully Backordered (Increased Production Rate)

A ($)	E₁*($)
900	8174.95
890	8161.50
880	8147.97
870	8134.36
860	8120.68
850	8106.91

I.3.2 CASE WHEN THE SHORTAGES ARE NOT FULLY BACKORDERED

As mentioned before, the total costs (Sharma 2007) can be transformed for the current scenario also. With the similar procedure, the following total cost can be used for relevant calculations:

$$E^* = (CD / P) + \sqrt{2DA(1 - D / P)H(K - Cb / P) / (K + H - Cb / P)} \qquad (13)$$

where b = proportion of shortages which is not backordered.

With the use of previous example, consider b = 0.2 for an upward production rate variation.

The costs are computed as follows:

E* = $8231.26
E₁* = $8123.57

Table I.4 shows the decrease in cost with a reduction in setup cost.

As expected, cost reduction is more sensitive with reduction in setup cost in this case also.

TABLE I.4
Revised Cost When Shortages Are Not Fully Backordered (Increased Production Rate)

A ($)	E₁*($)
900	8123.57
890	8110.40
880	8097.16
870	8083.84
860	8070.44
850	8056.97

I.4   CONCLUSION

After showing the relevance of appropriate batch size in a manufacturing concern, an increase in production rate is analyzed. Because of its possibility of being environment friendly, a case is explored where 'green' facility setup cost is quantified. An upper bound is obtained, below which the implementation becomes economical also. There is an incentive for the companies to decrease their setup cost below this threshold further, because the total cost is found to be more sensitive toward such setup cost decrease.

Since setup cost level and production rate level might differ from company to company, an analysis is extended for the manufacturing rate decrease also. A lower bound is obtained for setup cost, above which the implementation becomes economical. If a company has their setup cost much above this threshold level, it can find certain incentive for reducing the setup cost in this case too. A practical implementation procedure is also briefly discussed in the chapter for both upward as well as downward manufacturing rate variation.

REFERENCES

Sarker, B. R. and Babu, P. S., 1993, Effect of production cost on shelf life. *International Journal of Production Research*, Vol. 31, No. 8, 1865–1872.

Sharma, S., 2004, Optimal production policy with shelf life including shortages. *Journal of the Operational Research Society*, Vol. 55, No. 8, 902–909.

Sharma, S., 2006, Narrowing the gap between continuous and intermittent manufacturing. *Journal of Engineering Manufacture*, Vol. 220, No. 11, 1901–1905.

Sharma, S., 2007, Minimizing the difference between intermittent and continuous production with backorders. *Journal of Engineering Manufacture*, Vol. 221, No. 11, 1617–1623.

Sharma, S., 2008, On the flexibility of demand and production rate. *European Journal of Operational Research*, Vol. 190, No. 2, 557–561.

Sharma, S., 2009, Revisiting the shelf life constrained multi-product manufacturing problem. *European Journal of Operational Research*, Vol. 193, No. 1, 129–139.

Sharma, S., 2010, Policies concerning decisions related to quality level. *International Journal of Production Economics*, Vol. 125, No. 1, 146–152.

Silver, E. A., 1990, Deliberately slowing down output in a family production context. *International Journal of Production Research*, Vol. 28, No. 1, 17–27.

Silver, E. A., 1995, Dealing with a shelf life constraint in cyclic scheduling by adjusting both cycle time and production rate. *International Journal of Production Research*, Vol. 33, No. 3, 623–629.

Advanced Reading II
Toward a Synergy of the Theory of Exchange and Sanjay Sharma's Index for Relative Forecasting Expenditure

II.1 INTRODUCTION

In case of the single-/multi-item procurement, the following cost components are relevant:

 i. Purchase cost
 ii. Ordering cost
 iii. Inventory carrying cost

However, usually the ordering and inventory carrying costs are balanced in most of the cases along with the purchase cost in few cases. Recently, the author attempted an explicit inclusion of the cost related to forecast (Sharma 2016) in single-item procurement and production environment and approached a unique index for relative forecasting expenditure. For the production environment, shortage cost may also be considered additionally depending on the case along with the unit production cost. This chapter transforms the formulation and discussion for the multi-item environment.

In a separate research stream, the author swapped the production rates of two items in the multi-item environment (Sharma 2007a). It is further extended (Sharma 2007b). Inventory carrying/holding and shortage costs are also exchanged (Sharma 2007c). A methodology has been explained in which the possibility of exchange of demands of two items is explored (Sharma 2009a). The exclusive approach considering different parameters has also been summarized by the author as the 'theory of exchange' (Sharma 2008). This is also extended (Sharma 2009b, 2009c) for the single/multiple parameter swapping. But in this research/practice stream, a forecasting expenditure has not been incorporated. An obvious research direction is the synergy of these two aspects and the present chapter is devoted to this. The approach is also illustrated by Figure II.1.

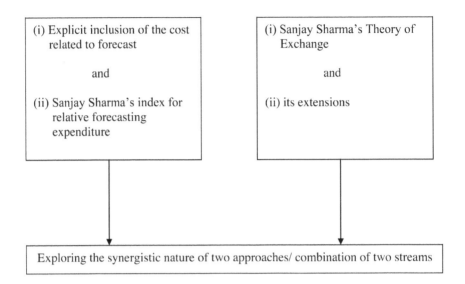

FIGURE II.1   Synergy of two approaches.

The present chapter is distinct in certain ways such as:

i. 'Theory-of-exchange'-related work considers a multi-item manufacturing environment. However, the forecasting expenditure is also relevant for the procurement scenario. Therefore, it is reasonable to start from the fundamental procurement case, and the formulation and discussion are transformed for this case also.

ii. 'Explicit forecasting expenditure'-related work considers single item since the objective was the conceptual understanding of the approach and the unique index. However, the companies are engaged in the multi-item procurement and production. Therefore, the formulation and analysis need adjustment to incorporate this scenario. The analysis includes the exchange of parameters.

The methodology is briefly explained for the following cases:

i. Procurement
ii. Production
iii. Complete backlogging
iv. Partial backlogging

II.2   FUNDAMENTAL PROCUREMENT CASE

To formulate and discuss, the following notations for an item i have been considered:

A_i = Ordering/setup cost for an item i
b_i = A fraction of shortage quantity which is not backlogged

c_i = Unit procurement/production cost

c_{1i} = A constant factor associated with the forecasting expenditure function

c_{2i} = A constant factor associated with any potential benefit obtained because of the increased forecasting expenditure

D_i = Annual demand for an item i

H_i = Annual inventory carrying cost per unit item

I = Sanjay Sharma's index for relative forecasting expenditure

K_i = Annual shortage cost per unit

p_i = Fractional decrease in the variation of actual demand from the forecast demand

q_i = Fractional increase in the forecasting expenditure

Conventionally, the total cost (E) is expressed in terms of the procurement cycle time (T) for n number of items as follows:

$$E = \frac{T}{2} \sum_{i=1}^{n} D_i H_i + \frac{1}{T} \sum_{i=1}^{n} A_i \qquad (1)$$

With the optimal value of cycle time,

$$T = \sqrt{\frac{2 \sum_{i=1}^{n} A_i}{\sum_{i=1}^{n} D_i H_i}}$$

the optimal cost is expressed as:

$$E^* = \sqrt{2 \left[\sum_{i=1}^{n} D_i H_i \right] \left[\sum_{i=1}^{n} A_i \right]} \qquad (2)$$

Formulation of the relevant cost incorporating the efforts related to forecast has been explained in detail for single item (Sharma 2016). Accordingly for several items,

$$E_1 = \frac{T}{2} \sum_{i=1}^{n} D_i H_i + \frac{1}{T} \sum_{i=1}^{n} (A_i + c_{1i} q_i) - \sum_{i=1}^{n} p_i D_i c_{2i} \qquad (3)$$

With the optimal cycle time,

$$T = \sqrt{\frac{2 \sum_{i=1}^{n} (A_i + c_{1i} q_i)}{\sum_{i=1}^{n} D_i H_i}}$$

the optimal cost:

$$E_1^* = \sqrt{2\left[\sum_{i=1}^{n} D_i H_i\right]\left[\sum_{i=1}^{n} (A_i + c_{1i}q_i)\right]} - \sum_{i=1}^{n} p_i D_i c_{2i} \tag{4}$$

For any potential cost improvement,

$$E_1^* - E^* < 0$$

Using the Eqs. (2) and (4),

$$\frac{\sqrt{2\left[\sum_{i=1}^{n} D_i H_i\right]}\left[\sqrt{\sum_{i=1}^{n} (A_i + c_{1i}q_i)} - \sqrt{\sum_{i=1}^{n} A_i}\right]}{\sum_{i=1}^{n} p_i D_i c_{2i}} < 1 \tag{5}$$

For single item, the following index (i.e., without subscript i) is introduced as Sanjay Sharma's index for relative forecasting expenditure (Sharma 2016):

$$I = \frac{\sqrt{2DH}}{pDc_2}\left[\sqrt{(A + qc_1)} - \sqrt{A}\right] \tag{6}$$

For the multi-item procurement, the condition (5) should be satisfied for an overall cost improvement. The left hand side (LHS) of Eq. (5) can now be called as the combined index for forecasting expenditure. Lower value is preferred for any business environment. However, in practice, the Eq. (6) may also be used for each item and focus may be on the item for which the value of I is the lowest. Else, the Eq. (5) will provide the holistic view considering all items.

For example, consider the relevant data given in Table II.1.

An initial analysis using the index expressed by Eq. (6) may suggest the following:

i. Item 1:

$$I = 0.65$$

ii. Item 2:

$$I = 0.97$$

It may seem that given the opportunity, the priority should be given to the item 1. However, considering the overall system view, the combined index expressed by Eq. (5) is as follows:

0.77; that is also less than 1.

TABLE II.1
Parameters for the Items

	Item i	
	$i = 1$	$i = 2$
A_i	$40	$50
c_{1i}	$2000	$1000
q_i	0.1	0.2
D_i	300 units per year	400 units per year
H_i	$12 per unit	$8 per unit
p_i	0.04	0.03
c_{2i}	100	60
K_i	$100 per unit	$80 per unit
b_i	0.2	0.1
P_i	750 units per year	720 units per year
c_i	$30	$20

II.2.1 ALLOWABLE BACKLOGGING IN PROCUREMENT

When the backorders are allowed in the procurement scenario, the conventional cost is expressed as:

$$E = \frac{T}{2} \sum_{i=1}^{n} \frac{D_i H_i K_i}{(H_i + K_i)} + \frac{1}{T} \sum_{i=1}^{n} A_i \tag{7}$$

With the optimal value of CT,

$$T = \sqrt{\frac{2 \sum_{i=1}^{n} A_i}{\sum_{i=1}^{n} D_i H_i K_i / (H_i + K_i)}}$$

the optimal cost:

$$E^* = \sqrt{2 \left[\sum_{i=1}^{n} D_i H_i K_i / (H_i + K_i) \right] \left[\sum_{i=1}^{n} A_i \right]} \tag{8}$$

Incorporating the explicit cost related to the forecast,

$$E_1 = \frac{T}{2} \sum_{i=1}^{n} \frac{D_i H_i K_i}{(H_i + K_i)} + \frac{1}{T} \sum_{i=1}^{n} (A_i + c_{1i} q_i) - \sum_{i=1}^{n} p_i D_i c_{2i} \tag{9}$$

and,

$$E_1^* = \sqrt{2\left[\sum_{i=1}^{n} D_i H_i K_i / (H_i + K_i)\right]\left[\sum_{i=1}^{n}(A_i + c_{1i}q_i)\right] - \sum_{i=1}^{n} p_i D_i c_{2i}}$$ (10)

Following a similar procedure, it can be shown that:

$$\frac{\sqrt{2\left[\sum_{i=1}^{n} D_i H_i K_i / (H_i + K_i)\right]}\left[\sqrt{\sum_{i=1}^{n}(A_i + c_{1i}q_i)} - \sqrt{\sum_{i=1}^{n} A_i}\right]}{\sum_{i=1}^{n} p_i D_i c_{2i}} < 1$$ (11)

Considering the relevant data from Table II.1, the combined index is obtained as follows:

0.729; which is less than the previous scenario (i.e., 0.77).

The objective of the present chapter is not to discuss/implement the author's theory of exchange in detail. However, as an example, consider the exchange of the shortage cost (Sharma 2007c). With the exchange of the shortage costs of two items, j and k in a group of n items,

$$E_1^* = \sqrt{\left[2\left[\sum_{\substack{i=1 \\ i \neq j \\ i \neq k}}^{n}\{D_i H_i K_i / (H_i + K_i)\} + D_j H_j K_k / (H_j + K_k) + D_k H_k K_j / (H_k + K_j)\right]\right]\left[\sum_{i=1}^{n}(A_i + c_{1i}q_i)\right]} - \sum_{i=1}^{n} p_i D_i c_{2i}$$ (12)

Difference between the right hand side (RHS) of Eqs. (10) and (12) should be greater than zero in order to get any potential benefit from this approach, i.e.,

$$\sqrt{2\left[\sum_{i=1}^{n}(A_i + c_{1i}q_i)\right]}\left[\sqrt{\sum_{\substack{i=1 \\ i \neq j \\ i \neq k}}^{n}\{D_i H_i K_i / (H_i + K_i)\} + D_j H_j K_j / (H_j + K_j) + D_k H_k K_k / (H_k + K_k)} - \sqrt{\sum_{\substack{i=1 \\ i \neq j \\ i \neq k}}^{n}\{D_i H_i K_i / (H_i + K_i)\} + D_j H_j K_k / (H_j + K_k) + D_k H_k K_j / (H_k + K_j)}\right]$$

> 0

The above condition is satisfied when:

$$\frac{D_j H_j K_j}{(H_j + K_j)} + \frac{D_k H_k K_k}{(H_k + K_k)} > \frac{D_j H_j K_k}{(H_j + K_k)} + \frac{D_k H_k K_j}{(H_k + K_j)} \tag{13}$$

Certain insights can also be obtained from the expression (11). The lower value for the LHS of (11) is preferred. Therefore, after an exchange of the shortage costs of the two items, the value of

$$\left[\sum_{i=1}^{n} D_i H_i K_i \, / \, (H_i + K_i) \right]$$

should be lower than before. The condition (13) relates to this aspect.

Using the relevant data from Table II.1,

$$\left[\sum_{i=1}^{n} D_i H_i K_i \, / \, (H_i + K_i) \right]$$

is obtained as follows:

  i. before: 6123.38;
  ii. after the exchange: 6093.40

In case where a fraction b_i of the shortages is not backlogged, conventionally:

$$E = \frac{T}{2} \sum_{i=1}^{n} \frac{D_i H_i (K_i - c_i b_i)}{(H_i + K_i - c_i b_i)} + \sum_{i=1}^{n} c_i D_i + \frac{1}{T} \sum_{i=1}^{n} A_i \tag{14}$$

As discussed before,

$$E_1 = \frac{T}{2} \sum_{i=1}^{n} \frac{D_i H_i (K_i - c_i b_i)}{(H_i + K_i - c_i b_i)} + \sum_{i=1}^{n} c_i D_i + \frac{1}{T} \sum_{i=1}^{n} (A_i + c_{1i} q_i) - \sum_{i=1}^{n} p_i D_i c_{2i} \tag{15}$$

And it can be shown that:

$$\frac{\sqrt{2 \left[\sum_{i=1}^{n} D_i H_i (K_i - c_i b_i) / (H_i + K_i - c_i b_i) \right]} \left[\sqrt{\sum_{i=1}^{n} (A_i + c_{1i} q_i)} - \sqrt{\sum_{i=1}^{n} A_i} \right]}{\sum_{i=1}^{n} p_i D_i c_{2i}} < 1 \tag{16}$$

With the relevant data of Table II.1, the combined index is calculated as:

0.727; which is lower than the previous scenario (i.e., 0.729).

For the implementation of the 'theory of exchange', the combined index expressed by (16) is useful. In case of an exchange of the shortage cost, as discussed before, the value of

$$\left[\sum_{i=1}^{n} D_i H_i (K_i - c_i b_i) / (H_i + K_i - c_i b_i)\right]$$

should be lower in comparison with the original situation.

II.2.2 FIXED CYCLE TIME

In some cases, it may not be possible to change the CT. The procurement CT is being practiced by the buyer as well as the supplier company for a long time and either one company or both may not be ready to change it.

With reference to the Eqs. (1) and (3),

$$(E_1 - E) < 0$$

shows:

$$\frac{1}{T} \sum_{i=1}^{n} c_{1i} q_i - \sum_{i=1}^{n} p_i D_i c_{2i} < 0$$

or,

$$T > \frac{\sum_{i=1}^{n} c_{1i} q_i}{\sum_{i=1}^{n} p_i D_i c_{2i}} \tag{17}$$

Since many companies focus on their CT, the condition, i.e., the CT should be greater than the ratio of periodic suggested increase in the forecasting expenditure and the annual estimated relevant benefit, is expected to be useful in the decision support.

II.3 PRODUCTION CASE

Consider the following additional notation for this case:

P_i = Annual rate of manufacture/production for an item i.

A detailed procedure has been explained before to find out the combined index. For a ready reference, this is presented briefly concerning the production scenario.

Using the following equation as reference,

$$E_1 = \frac{T}{2} \sum_{i=1}^{n} D_i H_i (1 - D_i / P_i) + \frac{1}{T} \sum_{i=1}^{n} (A_i + c_{1i} q_i) - \sum_{i=1}^{n} p_i D_i c_{2i} \tag{18}$$

It can be shown that:

$$\frac{\sqrt{2\left[\sum_{i=1}^{n} D_i H_i (1 - D_i / P_i)\right]}\left[\sqrt{\sum_{i=1}^{n}(A_i + c_{1i}q_i)} - \sqrt{\sum_{i=1}^{n} A_i}\right]}{\sum_{i=1}^{n} p_i D_i c_{2i}} < 1 \tag{19}$$

Considering the relevant data from Table II.1, the combined index from the above expression is computed as:

0.558

In certain cases, the organization may be interested in the comparison of standard costing and historical costing. The standard costing refers to the planned cost concerning the activities. Since it is the planned one, it should be available before the start of the period. On the other hand, the historical or actual costing is available at the end of the period. Standard costing is obviously on the basis of the estimated parameters/factors, whereas the historical costing refers to the actual expenditure on activities. The organization would like to analyze the variation between the standard and historical costing.

In the present multi-item production context also, the total relevant cost may be estimated before the start of period and may be compared with the actual total relevant cost at the end of the period. Figure II.2 represents this comparison with the following abbreviation along with their meaning:

t = period, i.e., 1, 2, 3, …
SC_t = standard costing for period t
HC_t = historical costing for period t

The end of the period t marks the beginning of the period $(t + 1)$. If the HC_t is more than the SC_t, then the detailed exploration is necessary. Since the SC_t depends on the estimation of parameters, the parameter values may need certain modification. As the forecasting expenditure function and the associated potential benefit have been discussed in the present chapter, the comparison procedure is expected to be useful in a revision of the following estimates also, among other conventional parameters:

i. c_{1i} and also the appropriate function for forecasting expenditure.
ii. c_{2i} and also the associated benefit owing to the planned, increased forecasting expenditure.

If the HC_t is less than the SC_t, the parameter/function values also need suitable revision.

The discussed iterative procedure is also illustrated by Figure II.3.

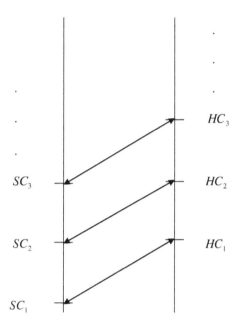

FIGURE II.2 Comparison of standard and historical costing.

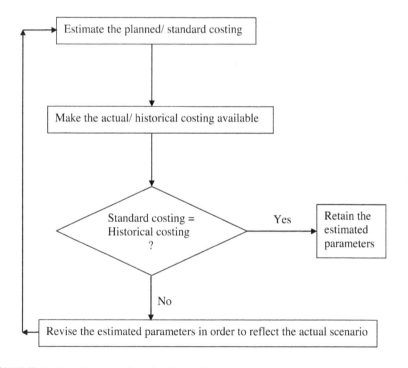

FIGURE II.3 Iterative procedure for the costing.

To implement the author's theory of exchange, the organization has the following options:

i. After obtaining a stable standard costing, implement the theory of exchange.
ii. An effort may be made to apply the theory of exchange without waiting for the stable planned costing.

If the period is relatively short, the company may wait for the standard costing to stabilize. After obtaining a suitable optimal level in terms of the cost, an attempt may be made to apply the theory of exchange in order to decrease the relevant costs further.

If the period is relatively long, then it may not be advisable to wait for the stability of the standard costing. An implementation of the theory of exchange is possible in parallel. In either case, depending on the scenario, however, once a certain level of an estimated relevant cost is available, the swapping of the suitable parameters may be considered in order to gain further improvement in the cost.

II.3.1 ALLOWABLE BACKLOGGING IN PRODUCTION

For this case,

$$E_1 = \frac{T}{2} \sum_{i=1}^{n} \frac{D_i H_i K_i (1 - D_i / P_i)}{(H_i + K_i)} + \frac{1}{T} \sum_{i=1}^{n} (A_i + c_{1i} q_i) - \sum_{i=1}^{n} p_i D_i c_{2i} \qquad (20)$$

and

$$\frac{\sqrt{2 \left[\sum_{i=1}^{n} D_i H_i K_i (1 - D_i / P_i) / (H_i + K_i) \right] \left[\sum_{i=1}^{n} (A_i + c_{1i} q_i) - \sqrt{\sum_{i=1}^{n} A_i} \right]}}{\sum_{i=1}^{n} p_i D_i c_{2i}} < 1 \qquad (21)$$

When a fraction b_i of the shortage quantity is not backordered,

$$E_1 = \frac{T}{2} \sum_{i=1}^{n} \frac{D_i H_i (K_i - c_i b_i)(1 - D_i / P_i)}{(H_i + K_i - c_i b_i)} + \sum_{i=1}^{n} c_i D_i + \frac{1}{T} \sum_{i=1}^{n} (A_i + c_{1i} q_i) - \sum_{i=1}^{n} p_i D_i c_{2i}$$

$$(22)$$

and

$$\frac{\sqrt{2 \left[\sum_{i=1}^{n} D_i H_i (K_i - c_i b_i)(1 - D_i / P_i) / (H_i + K_i - c_i b_i) \right] \left[\sum_{i=1}^{n} (A_i + c_{1i} q_i) - \sqrt{\sum_{i=1}^{n} A_i} \right]}}{\sum_{i=1}^{n} p_i D_i c_{2i}} < 1$$

$$(23)$$

II.4 DISCUSSION FOR THE CASE OF SINGLE STRATEGIC ITEM

In the Section 2, the index is obtained individually for each independent item along with the combined index. However, in a multi-item system environment, the focus on one item affects the system cost. Furthermore, the case exists where the company wants to enhance the forecasting efforts (i.e., the cost) related to one strategic item. After finding out the strategic item with the use of individual index among other business aspects, an additional analysis is necessary in the system context.

II.4.1 PROCUREMENT

In a group of n items ($i= 1,2,...n$), assume that j is a strategic item. With reference to the Eq. (4), now:

$$E_1^* = \sqrt{2\left[\sum_{i=1}^{n} D_i H_i\right]\left[c_{1j}q_j + \sum_{i=1}^{n} A_i\right]} - p_j D_j c_{2j} \tag{24}$$

Using the Eq. (2) and (24),

$$E_1^* - E^* < 0$$

shows:

$$\frac{\sqrt{2\left[\sum_{i=1}^{n} D_i H_i\right]}\left[\sqrt{(c_{1j}q_j + \sum_{i=1}^{n} A_i)} - \sqrt{\sum_{i=1}^{n} A_i}\right]}{p_j D_j c_{2j}} < 1 \tag{25}$$

Considering the related parameters from Table II.1, the index is obtained for j=1 as follows:

0.733; which is less than 1.

For j=2, this index is 1.61; the condition (25) is not satisfied as it is greater than 1.

Following a similar procedure, the related index can be obtained for other situations as follows:

i. Complete backlogging:

$$\frac{\sqrt{2\left[\sum_{i=1}^{n} D_i H_i K_i / (H_i + K_i)\right]}\left[\sqrt{(c_{1j}q_j + \sum_{i=1}^{n} A_i)} - \sqrt{\sum_{i=1}^{n} A_i}\right]}{p_j D_j c_{2j}}$$

ii. Partial backlogging:

$$\frac{\sqrt{2\left[\sum_{i=1}^{n} D_i H_i (K_i - c_i b_i) / (H_i + K_i - c_i b_i)\right]}\left[\sqrt{(c_{1j}q_j + \sum_{i=1}^{n} A_i)} - \sqrt{\sum_{i=1}^{n} A_i}\right]}{p_j D_j c_{2j}}$$

II.4.2 PRODUCTION

As explained before, the index is as follows:

$$\frac{\sqrt{2\left[\sum_{i=1}^{n} D_i H_i (1 - D_i / P_i)\right]}\left[\sqrt{(c_{1j}q_j + \sum_{i=1}^{n} A_i)} - \sqrt{\sum_{i=1}^{n} A_i}\right]}{p_j D_j c_{2j}}$$

After incorporating the shortages, the index for each case is provided below:

i. Complete backlogging:

$$\frac{\sqrt{2\left[\sum_{i=1}^{n} D_i H_i K_i (1 - D_i / P_i) / (H_i + K_i)\right]}\left[\sqrt{(c_{1j}q_j + \sum_{i=1}^{n} A_i)} - \sqrt{\sum_{i=1}^{n} A_i}\right]}{p_j D_j c_{2j}}$$

ii. Partial backlogging:

$$\frac{\sqrt{2\left[\sum_{i=1}^{n} D_i H_i (K_i - c_i b_i)(1 - D_i / P_i) / (H_i + K_i - c_i b_i)\right]}\left[\sqrt{(c_{1j}q_j + \sum_{i=1}^{n} A_i)} - \sqrt{\sum_{i=1}^{n} A_i}\right]}{p_j D_j c_{2j}}$$

In addition to the selection of the strategic item, it is also relevant to choose suitable parameters for exchange. The next section deals with the selection of parameters for exchange/swapping.

II.5 SELECTION OF PARAMETERS FOR SWAPPING

In the manufacturing/production scenario, various items are produced on either a machine or a cluster of machines. Following parameters are relevant in the context of the basic production case:

i. Manufacturing time cost
ii. Manufacturing CT
iii. Annual inventory holding cost fraction
iv. Rate of demand
v. Rate of manufacture

Among these parameters, manufacturing time cost depends on the duration in which the facility is active and is usually constant for a planning period. Annual inventory holding cost fraction is also treated as uniform. However, the inventory holding cost depends on the item concerned. This has a bearing on the procurement cost and subsequent value addition. The CT is an output parameter obtained after optimizing the total relevant cost.

Rate of demand is conventionally an input parameter. There are situations in which the management is interested in substituting the existing higher demand of an old item with the presently lower demand of a recent innovative item having a similar application. The idea may be to phase out the old item in the long run, and it is necessary to know the effect of this transition on the total cost. Otherwise also, it is of interest to implement the swapping of the demands of two products in case where this provides certain cost benefit.

Rate of manufacture is a function of available options/range concerning the facility. In many cases, it depends on the skill, qualifications, and experience of the human resources associated with the operation. In a truly flexible system, it is not difficult to vary the rate of manufacture. Otherwise also, in the conventional continuous production, usually similar item is produced in a wide variety of specifications and therefore may be treated as the separate items for scheduling purposes. For instance, on a tube mill, the steel tubes are made with different specifications related to:

     i. Diameter
     ii. Thickness
     iii. Length
     iv. Metallurgical properties

In an industry, the manufacturing facility is run conventionally in an optimal manner. Unless a conscious effort is made to swap the rate of manufacture of two items, it is not possible to visualize the benefit. It should be implemented in case of the cost improvement.

In the previous section, the case for single strategic item has been analyzed. Depending on the organizational structure, the following factors also play a significant role in selecting a strategic item for the swapping purpose:

     i. Human intensive operation
     ii. Flexibility incorporated in the system
     iii. Profit contribution of the item
     iv. Marketing strategy

Else in order to begin with, any item may be selected at random and the methodology is applied. After gaining some experience, further refinements can be made.

II.6 CONCLUSION

Recently, the author included the forecasting expenditure explicitly in the single-item procurement and production cases. This chapter has transformed the analysis for multi-item cases. After the development of combined indices, these have been shown to be useful for an implementation of the author's theory of exchange. Wherever necessary, the synergy is illustrated briefly with the appropriate example.

The combined index is lower for production scenario as compared to the procurement case. Therefore, given the choice, a priority can be allowed to the production case in the context of certain forecasting expenditure.

In some cases, the focus of a company may be on one strategic item in a group. The indices are presented to enhance the usefulness of the developed approach in this practical environment. Certain issues are also discussed in this chapter concerning the selection of few significant parameters along with the selection of strategic item in the context of the organization. Finally, the approach and indices for different cases developed in this chapter are expected to be helpful directly in the decision support systems of the said environment.

REFERENCES

Sharma, S., 2007a, A fresh approach to performance evaluation in a multi-item production scenario. *European Journal of Operational Research*, Vol. 178, No. 2, 627–630.

Sharma, S., 2007b, A procedure for benchmarking in multi-product manufacturing. Proc IMechE. *Part B: Journal of Engineering Manufacture*, Vol. 221, No. 3, 541–546.

Sharma, S., 2007c, Interchange of the holding/shortage costs in multiproduct manufacture. Proc IMechE. *Part B: Journal of Engineering Manufacture*, Vol. 221, No. 1, 135–140.

Sharma, S., 2008, Theory of exchange. *European Journal of Operational Research*, Vol. 186, No. 1, 128–136.

Sharma, S., 2009a, A method to exchange the demand of products for cost improvement. *International Journal of Advanced Manufacturing Technology*, Vol. 45, No. 3–4, 382–388.

Sharma, S., 2009b, Extending Sanjay Sharma's theory of exchange. *International Journal of Applied Management Science*, Vol. 1, No. 4, 325–339.

Sharma, S., 2009c, Single/multiple parameter swapping in the context of Sanjay Sharma's theory of exchange. *International Journal of Advanced Manufacturing Technology*, Vol. 40, No. 5–6, 629–635.

Sharma, S., 2016, A unique index with the explicit cost related to forecast. *Journal of Modelling in Management*, Vol. 11, No. 2, 518–535.

Index